Charge Density and Structural Characterization of Thermoelectric Materials

R. Saravanan

Thermoelectric materials permit the direct conversion of temperature differences into electric energy, and vice versa. They are therefore of highest technological interest in applications such as solid state coolers, waste heat recovery, sensors and detectors, and power generators including remote power generation.

Thermoelectric materials are often called "environmentally green", and for good reasons. Not only can they help generate electrical energy from waste gases as they are generated in such processes as home heating, industrial fabrication and automotive motion. In cooling applications they eliminate the use of chemical refrigerant gases. Moreover, as thermoelectric conversion devices have no moving parts, they operate silently and have a very long life expectancy. The only current drawback of these devices is their poor efficiency.

Scientists all over the world are therefore studying the structural, thermoelectric, charge-density and magnetic properties of the most promising types of these materials at the atomic and electronic level. In addition to providing an introduction to the field, the main objective of this book is to present the results of the growth and structural characterization of thermoelectric materials that are of high current interest; including Mg_2Si, PbTe, $Bi_{1-x}Sb_x$, Bi_2Te_3, Sb_2Te_3, $Sn_{1-x}Ge_xTe$ and InSb.

Charge Density and Structural Characterization of Thermoelectric Materials

Dr. R. Saravanan, M.Sc., M.Phil., Ph.D.
Associate Professor & Head
Research Centre and PG Department of Physics
The Madura College (Autonomous)
Madurai - 625 011

Published by **Materials Research Forum LLC**
Millersville, PA 17551, USA

Published as part of the book series
Materials Research Foundations
Volume 1 (2016)

ISSN 2471-8890 (Print)
ISSN 2471-8904 (Online)

ISBN 978-1-945291-00-5 (Print)
ISBN 978-1-945291-01-2 (eBook)

Distributed worldwide by

Materials Research Forum LLC
105 Springdale Lane
Millersville, PA 17551
USA
http://www.mrforum.com/

Manufactured in the United State of America
10 9 8 7 6 5 4 3 2 1

Table of Contents

Preface

The study of materials in terms of their effectiveness/usefulness in device applications is important, for the technological development of a country. In this sense, thermoelectric materials have their own standing in their device applications, such as solid state coolers, power generators including remote power generation, waste heat recovery, sensors and detectors. The devices based on thermoelectric property have various advantages over other techniques, like long life period, silent operation, no moving parts, no chemical refrigerant gases and they can be considered "environmentally green". Their only current drawback is poor efficiency. The scientific and technological task of achieving good thermoelectric materials is difficult because these materials need to have high electrical conductivity, high thermoelectric power and low thermal conductivity (Phonon Glass Electron Crystal). Microscopic analyses (electron level properties) of such materials are necessary for the construction of more efficient thermoelectric devices.

The electron level properties of thermoelectric materials can be analyzed using many techniques. X-ray diffraction analysis is one of the techniques, which gives finite details of the materials in the form of average structure, charge density and local structure. A qualitative picture of the bonding in atoms and quantitative charge density will provide a better understanding of the physical and chemical properties of thermoelectric materials. The inter-atomic interaction studies (local structure) along with the bonding will give better confirmation of the electron level properties, which enhances the thermoelectric behavior.

The materials discussed for the present book are the conventional bulk thermoelectric materials, Mg_2Si, PbTe, $Bi_{1-x}Sb_x$, Bi_2Te_3, Sb_2Te_3, $Sn_{1-x}Ge_xTe$ and InSb. The materials Mg_2Si, PbTe, $Bi_{1-x}Sb_x$, Bi_2Te_3 and Sb_2Te_3 were studied using powder X-ray diffraction data sets and the results are presented. The materials $Sn_{1-x}Ge_xTe$ and InSb were studied using single crystal X-ray data sets.

The main objective of this book is to present the results of the growth and structural characterization of these thermoelectric materials. A superior thermoelectric material for device applications such as solid state coolers, power generators, sensors and detectors is a material with high electrical conductivity, high thermoelectric power and low thermal conductivity in the same material (Phonon Glass Electron Crystal).

In most of the available national and international literatures of this field, only the growth and physical characterization have been given importance but not on the complete structural analysis. The physical characterization is important, however, the average and local structure and the elucidation of the electron density distribution with respect to concentration of dopant atoms will give one a complete analysis and a full understanding of the thermoelectric materials, which is most significent for current projects.

Hence, the present study is aimed to give fruitful information in the advancement of knowledge not only in the physical sense but also on the complete structural analysis in terms of relevant structural parameters like thermal vibration, atomic displacements, bond lengths and electron density etc., for both single crystals and powder samples. The structural information like bonding and electron density distribution were analyzed in this work using the versatile technique called maximum entropy method (MEM). The three dimensional and two dimensional pictorial visualization of the bonding between the atoms in terms of the electron density distribution in relevant plane and particular direction inside the unit cell is another milestone in the study of crystallographic materials. The nearest neighbor interactions were elucidated using the advanced technique called atomic pair distribution function (PDF). It is worthy to note that the above said structural information was elucidated from the refined structure factors based on the least square profile fitting methodology. All the above said structural information was extracted from powder and single crystal X-ray intensity data sets. In brief, the main objectives of the book are as follows.

Chapter I gives the origin, history, theoretical background and the applications of thermoelectric materials. This chapter also deals with the basic theory of X-ray diffraction analysis, powder X-ray method and crystal structure determination from powder diffraction data sets. The methodology of least square refinement using Rietveld technique and its strategies are discussed. The algorithm and detailed procedure of charge density estimation models like MEM calculations with refinement strategies are also presented. The basic theory of local and average structure analysis by pair distribution function (PDF) has been discussed and the detailed methodology adapted in this work is also presented in this chapter. The earlier works on thermoelectric materials selected for our study, Mg_2Si, PbTe, $Bi_{1-x}Sb_x$, Bi_2Te_3, Sb_2Te_3, $Sn_{1-x}Ge_xTe$ and InSb and the schematic outline of the present work are presented in this chapter.

Chapter II deals with the results and discussion of the average structural analysis of the selected thermoelectric materials using least square refinement method, employing Rietveld technique. The fitted XRD profiles for the selected thermoelectric materials, for which the powder X-ray diffraction data is taken, are plotted. The numerical results of the structure factor, both observed and calculated are tabulated. This chapter also reveals the discussion of the thermoelectric properties of the selected materials based on the Rietveld analysis, particularly in terms of cell parameters, thermal vibration parameters.

Chapter III gives the results and discussion of the charge density analysis maximum entropy method (MEM). The 3D and 2D pictorial results of the selected materials in various planes are plotted and the bonding between the atoms in different planes is discussed. The thermoelectric phenomenon is discussed based on the nature of the bonding and its numerical charge density values.

Chapter IV deals with the study of local structure *ie.,* the inter-atomic ordering of the selected thermoelectric materials in terms of the first, second and third coordination shells. The pair distribution function (PDF), atomic correlation function is computed for the selected thermoelectric materials. The numerical results of the first, second and third inter-atomic distances are computed and compared with the MEM results.

Chapter V presents the comprehensive conclusion of the results of the reported work along with a list of publications. The citations of the published papers by different authors are also listed here.

Some of the results of the present work have been previously published by the author of this book in reputed journals as follows:

1. Local structure of the thermoelectric material Mg_2Si using XRD, R. Saravanana, M. Charles Robert - Journal of Alloys and Compounds 479 (2009) pp. 26–31

2. Triple phase structure and electron density analysis of the thermoelectric material $Bi_{80}Sb_{20}$, M. Charles Robert, R. Saravanan - Powder Technology 197 (2010) pp. 159–164

3. Local structure of the high-temperature thermoelectric material PbTe using the maximum entropy method (MEM) and pair distribution function (PDF), R. Saravanan, M. Charles Robert, Journal of Physics and Chemistry of Solids 70 (2009) pp.159–163

4. Single Crystal Charge Density Studies of Thermoelectric Material Indium Antimonide, M. Charles Robert, B.Subha, R.Saravanan, Z. Naturforsch. 66a, 562–568 (2011) / DOI: 10.5560/ZNA.2011-0004

5. Single crystal X-ray analysis of the electronic structure of the thermoelectric material $Sn_{1-x}Ge_xTe$, M. Charles Robert, R. Saravanan - *Indian J. Phys.* 84 (9) (2010) pp. 1203-1210

6. Experimental electronic structure of the thermoelectric materials Bi_2Te_3 and Sb_2Te_3, T. Akilan, M. Charles Robert, R. Saravanan, *Materials Science Forum Vol. 699 (2012) pp 103-121*

CHAPTER I

Thermoelectrics: An Introduction

Abstract

The thermoelectric phenomenon is a direct energy conversion process by electrons in solids and this phenomenon has various advantages in harmony with our environment. The devices based on thermoelectric property have various advantages over other techniques and are "environmentally green". The thermoelectric mechanism is not macroscopic, but depends on the microscopic phenomena such as crystal structure, nature of the bonding between constituent atoms, lattice thermal vibration effects, electron density distribution between the atoms and the local structural distribution between the nearest neighbor atoms. Chapter I reviews the origin, historical improvements, the basic theory of thermoelectric figure of merit, the current status of thermoelectrics, the structural refinement using least squares technique, the analysis of electron density distribution using currently available novel techniques and the real space analysis of the nearest neighbor distance using atomic correlation functions. The experimental and theoretical background and some reviews of the earlier works have also been presented in this chapter. Finally, a brief outline of the present book is given at the end of this chapter.

Keywords

Thermoelectric Mechanism, Thermoelectric Materials, Applications, Maximum Entropy Method, Pair Distribution Function, X-Ray Diffraction

Contents

1.1 THERMOELECTRICS: AN INTRODUCTION

The development of any country relies on the research and development, production in large scale and usage of advanced novel materials. In this context, the thermoelectric materials have their own stand because of their major device applications. Also, the important crises like global warming, ozone depletion and heavy oil shortage can be solved by the usage of effective thermoelectric materials. Global warming can effectively be reduced by the proper usage of thermoelectric devices based on thermoelectric phenomena, because these materials have the ability to convert waste heat energy into usable electrical energy [Rowe, 2006]. The applications include the conversion of waste heat into electricity in micro/nano integrated chips to large scale industries like steel plants. The ozone layer is depleted, because of the ozone holes partly produced by the chloro-fluoro carbons (CFC) emitted from the household refrigerators. The usage of CFC can be completely eliminated by using alternate refrigerators called thermoelectric coolers. Around 65% of fuel energy is wasted as heat in almost all diesel powered vehicles. This waste energy can be converted into useful electrical energy and stored in auxiliary batteries [Rowd, 2006; Yodovard *et al.*, 2001; Riffat, 2003]. Thus, the fuel efficiency can be enormously increased.

The thermoelectric phenomenon and technology have attracted a renewed interest from the viewpoint of increasing needs of environment-friendly energy sources. In the last decade, new thermoelectric materials have been searched extensively, some of which have better thermoelectric properties than the conventional thermoelectric materials. From the viewpoint of basic science, the thermoelectric power depends on entropy change mediated through electrons. This is more or less a controversial terminology because the entropy and heat are concepts in the macroscopic world, whereas the electron is a concept in the microscopic world. Thus, thermoelectric effect is laying a boundary between microscopic and macroscopic worlds, which will give a new insight or direction to condensed matter physics [Ichiro Terasaki, 2005].

1.2 ORIGIN

The Seebeck effect and Peltier effect are the most predominant thermoelectric effects. In 1822, Seebeck [Seebeck, 1822] observed that when two electrically conducting materials are connected in a closed loop with a temperature difference at the junctions T_1 and T_2, then there was a deflection of the magnetic needle in the measuring apparatus [Ioffe, 1957; Paul, 1960]. The deflection was dependent on the temperature difference between the two junctions and the materials used for the conduction. Shortly after this, Oersted [Brian and Cohen, 2007] discovered the interaction between an electric current and a magnetic needle. Many scientists subsequently researched the relationship between electric current and magnetic fields including Ampere, Biot, Savart and Laplace. Through these studies, it was discovered that the observation by Seebeck was not caused by a magnetic polarization, but was caused by electrical current flowing in the closed loop circuit. The electromotive force or voltage driving the electric current, can be measured by breaking the closed loop and measuring the open loop voltage ΔV, which is given by

$$\Delta V = \int_{T_1}^{T_2} S_{AB} dT \tag{1.1}$$

where S_{AB} is the Seebeck coefficient for the two conductors A and B, which is defined as being positive when a positive voltage is measured for $T_1 < T_2$.

While the Seebeck coefficient is associated with a couple formed by two materials, it was later discovered through the Thomson effect that an absolute Seebeck coefficient could be associated with each material individually as $S_{AB} = (S_A - S_B)$. Later, the absolute Seebeck coefficient for a given material is referred as the thermoelectric power of the material. Thermoelectric power will be positive for a material corresponding to a P type (hole assisted conduction) and negative for N type (electron assisted) conduction.

Twelve years after Seebeck's discovery, in the year 1834, a watchmaker and scientist named Jean Peltier reported a temperature anomaly at the junction of two dissimilar materials as a current was passed through the junction [Peltier, 1834]. It was unclear what caused this anomaly and Peltier attempted to explain it on the basis of the electrical conductivity and hardness of the two materials. Lenz removed all doubts in 1838 with one simple experiment. By placing a droplet of water in a dimple at the junction between rods of bismuth and antimony, Lenz was able to freeze the water and subsequently melt the ice by changing the direction of electric current through the junction. In this way, Lenz made the first thermoelectric cooler. The rate of heat (Q) absorbed or liberated from the junction was later found to be proportional to the current (I),

$$Q = \pi I \tag{1.2}$$

where, the proportionality constant π is named as the Peltier coefficient. This is the reverse process of Seebeck effect. According to Onsagar relation, S and π satisfy the relation [Onsagar, 1931] as

$$\pi = S_{AB} T \tag{1.3}$$

In order to get the maximum heat absorption or liberation, the junction should be formed of materials with thermoelectrical powers of opposite sign (i.e., one N type and another P type material).

1.3 THERMOELECTRIC MECHANISM

The ability to absorb or emit heat at the junction of the two materials led to the discovery of a useful temperature controlling device, through which in principle, it was possible to convert heat into electricity, due to the coupling between thermal and electrical phenomena. Such energy conversion technology is called thermoelectrics. Since the energy conversion is done by electrons in solids, firstly, the device can be made with no moving parts and can be operated without maintenance. Secondly, it produces no waste matter through conversion process and thirdly, it can be processed at a micro/nano size and can be implemented into electronic devices. This is a reversible process in which the electrical current can be supplied through the junction to create a temperature gradient (and heat flow), or a temperature gradient can be supplied to create electric current flow [Ioffe, 1957].

1.4 THERMO DYNAMICS OF THERMOELECTRIC DEVICES

1.4.1 INEQUILIBRIUM THERMO DYNAMICS

Quite generally, the electric current density j (particle flow) and the thermal current density q are written as functions of the gradient of chemical potential $\nabla\mu$ and the gradient of temperature $\nabla(1/T)$ as

$$-j = L_{11}\,\frac{1}{T}\,\nabla\mu + L_{12}\,\nabla \tag{1.4}$$

$$q = L_{21}\,\frac{1}{T}\,\nabla\mu + L_{22}\,\nabla\,\frac{1}{T} \tag{1.5}$$

Where L_y's are transport parameters [Callen, 1985]. The chemical potential consists of an electrostatic part $\mu_e = eV$ and a chemical part μ_c. Then the electric field is given as

$$E = -\nabla V = -\frac{1}{e}\,\nabla(-T)\,(\mu-\mu_c) \tag{1.6}$$

However, $\nabla\mu_c$ cannot be observed separately in real experiments, and is considered to be included in the observed E [Ashcroft and Mermin, 1976].

Then the above equations are identical to the Boltzmann transport equations given as

$$j = \sigma E + S\sigma\,\nabla(-T) \tag{1.7}$$

$$q = ST\sigma E + k'\nabla(-T) \tag{1.8}$$

where σ is the electrical conductivity and k' is the thermal conductivity for $j \neq 0$. Then, for $\nabla T = 0$, the electric field term can be eliminated from the equation and one can obtain

$$\frac{q}{T} = S\,j. \tag{1.9}$$

Since the left hand side is the entropy current density, the thermo power S is equivalent to the ratio of the entropy current to the electric current, or is equivalent to entropy per carrier.

1.4.2 FIGURE OF MERIT AND CONVERSION EFFICIENCY

In the cold side, the pumped heat Q_c is expressed as,

$$Q_C = ST_C I - \frac{1}{2} RI^2 - K\Delta T \tag{1.10}$$

The maximum heat absorbed by the cooling device can be calculated for constant T_h and T_c using a necessary condition $dQ_c/dI = 0$, which gives the optimum current $I_0 = ST_c/R$. By putting I_0 into eqn.1.10, gives,

$$Q_C^{max} = \frac{S^2T_c^2}{2R} - K\Delta T = \left(\frac{S^2T_c^2}{2RK} - \Delta T\right)K \tag{1.11}$$

Then, the figure of merit Z, can be defined as

$$Z = \frac{S^2}{RK} = \frac{S^2}{\rho K}$$

and rewrite Q_c^{max} as

$$Q_C^{max} = K\left(\frac{1}{2} ZT_c^2 - \Delta T\right) \tag{1.12}$$

Thus the maximum heat absorption is directly proportional to Z (the power factor) for $\Delta T = 0$.

Next, the lowest achievable temperature T_{co} for constant Q_c and T_H can be calculated using the necessary condition $dT_c/dI = 0$, which gives the optimum current

$$I_1 = \frac{ST_{CO}}{R}.$$

By putting I_1 into equation (1.10), gives,

$$\Delta T = \frac{S^2T_{CO}^2}{2KR} - \frac{Q_C}{K} = \frac{1}{2}ZT_{CO}^2 - \frac{Q_C}{K} \tag{1.13}$$

and the maximum temperature difference (*i.e* lowest achievable temperature) is again directly proportional to Z for Q_C =0.

The energy conversion efficiency for a cooling device is characterized by the coefficient of performance (COP) Φ, defined as

$$\Phi = \frac{Q_C}{W} = \frac{Q_C}{Q_H - Q_C} = \frac{ST_C I - RI^2/2 - K\Delta T}{(S\Delta T + RI)I} \quad (1.14)$$

Taking $d\Phi/dI = 0$, the optimized current I_2 as obtained is

$$I_2 = \frac{S\Delta T}{R(\sqrt{1+Z\overline{T}}-1)}$$

where, $\overline{T} = (T_C+T_H)/2$. By putting I_2 into Φ results

$$\Phi_{max} = \frac{T_C\sqrt{1+Z\overline{T}}-T_H}{\Delta T(\sqrt{1+Z\overline{T}}+1)} \quad (1.15)$$

For the power generation, the efficiency η is given as

$$\eta = \frac{W}{H} = \frac{IV}{ST_H I - \frac{RI^2}{2} + K\Delta T}$$

$$= \frac{x\Delta T}{(1+x)\overline{T} + \frac{(1+x)^2}{Z} + x\Delta T/2}$$

By taking $d\eta/dx = 0$, the maximum efficiency [Heikes and Ure, 1961] obtained is

$$\eta_{max} = \frac{\Delta T(\sqrt{Z\overline{T}+1}-1)}{T_H(\sqrt{Z\overline{T}+1}+\frac{T_C}{T_H})} \quad (1.16)$$

First Φ_{max} and η_{max} given by equation (1.15 and 1.16) are reduced to the Carnot efficiency as $ZT \rightarrow \alpha$. This is reasonable, because thermoelectric energy conversion is a conversion through the electron transport, which is an irreversible process accompanying the Joule heat. Secondly, the efficiency η is larger for larger ZT and ΔT. For practical purposes, the conversion efficiency η of 10-15 % (solar battery) is expected, which corresponds to $Z > 3\times10^{-3}\ K^{-1}$ with $\Delta T > 300K$. This means that ZT = 1.8 at 600 K is necessary. Thirdly COP of a commercial refrigerator is 1.2 to 1.3, which correspondents to ZT=4. Thus much improvement in ZT is needed to replace a thermoelectric cooler in the place of Freon gas refrigerator [Androulakis and Pantelis Migiakis, 2004].

1.5 THERMOELECTRIC MATERIALS

Metals typically have thermoelectric power (S) of the order of μV/K, which are too small for most practical applications with the exception of thermocouples. Many semiconductors, however, have much larger values of *S*, of the order of hundreds of μV/K. Although metals produce a smaller potential for a given temperature difference, they are good thermocouple materials because they are inexpensive and can easily be operated in high temperature environments.

Large thermo power values are important for a good thermoelectric material in addition to some of the other important factors. Since charge carriers must move through the material to transport heat, the material should conduct electricity well. Otherwise, the deleterious effect of resistive heating will be enhanced. In addition, the material should act as a thermal insulator. The purpose of the device (when operated as a heat pump) is to produce a hot and cold region, so a good thermal conductor will rapidly dissipate the temperature difference established. The best thermoelectric materials involve a trade-off among the three factors combining, a high thermo power and high electrical conductivity with low thermal conductivity. All three parameters are affected by the carrier concentration, *n*, of a solid [Ellis *et al.*, 1993]. Carrier concentrations range from about 10^{14} to 10^{21} carriers/cm^3 in a semiconductor and are about 10^{22} cm^{-3} in a metal. Electrical conductivity (σ) increases with the number of charge carriers (*n*). The greatest thermoelectric figure of merit (*Z*) value is obtained with a carrier concentration between 10^{18} and 10^{21} cm^{-3} [Mahan, 1998]. This implies that the best thermoelectric materials will be semiconductors with a relatively high carrier concentration.

The choice of carrier type is also important. The direction of both the Seebeck and Peltier effects is reversed depending on whether the carriers are electrons or holes. If both carrier types are present in a material, their effects will work against each other. Semiconductors always contain both carrier types, but often the semiconductor is intentionally laced with impurities (doped) so that one carrier type is greatly predominant [Ellis *et al.*, 1993]. In this case, the semiconductor is said to be extrinsic. Intrinsic semiconductors, on the other hand, have roughly equal numbers of each type of carrier, causing their performance as thermoelectric materials to suffer. Extrinsic semiconductors, then, are the better choice for thermoelectric devices. The existence of both positive and negative thermo-elements is of great utility in terms of device construction.

The next concept is the semiconducting energy gap. Good thermoelectric materials should have energy gaps of the order of 10 k_BT_{op}, where T_{op} is the operating temperature and k_B, the Boltzmann constant. Small gaps are generally good for thermoelectric performance because they lead to higher carrier mobility. However, if the gap is too small then the thermal excitation of minority carriers will adversely affect the figure of merit, since electrons and holes carry heat in opposite directions. A deep study of this has been given by Mahan *et al.,* [1989] saying that $10k_BT$ rule holds for direct and indirect gap semiconductors and for both phonon and impurity scattering.

The theory of thermoelectrics shows that $Z \propto \mu (m^*)^{3/2}$, where μ is the carrier mobility and the effective mass (m*) [Goldsmid, 1986]. Therefore it is desirable to maximize both m* and μ for good thermoelectric performance. Two important guidelines for materials result from the above proportionality. The first follows from the observation that m* can be increased without affecting μ much, if the semiconductor has several equivalent bands. Therefore good thermoelectrics are likely to be multi-valley semiconductors, and crystal structures with high symmetry with several equivalent bands [Mahan, 1998; Goldsmid, 1986]. The second concerns the electro-negativity difference between the elements making up the thermoelectric material [Slack, 1995]. The electro-negativity difference is a measure of the co-valency of the bonding in a material. Large electro-negativity differences indicate ionic bonding, large charge transfer, and strong scattering of electrons by optical phonons. This strong scattering leads to low carrier mobility, and is one of the reasons why oxides are generally poor thermoelectric materials. High electron mobility, on the other hand, is found in materials composed of elements with very similar values of electro-negativity. Good thermoelectrics, then, are composed of elements having small differences in electro-negativity.

The thermal conductivity (k) has two components, a lattice thermal conductivity (k_{lat}) and an electronic thermal conductivity (k_{ele}), such that $k = k_{lat} + k_{ele}$. The lattice component does not vary significantly with *n* (carrier concentration) but the electronic component increases with *n*. The thermo power, *Q*, generally decreases with further increase of carrier concentration. The figure of merit (ZT) will be the maximum in the boundary region between semiconductor and semi-metal.

Finding materials with exceptionally low values of lattice thermal conductivity is another concern about good thermoelectric material. A simple but useful expression for the lattice component of thermal conductivity is given by $k_{Lattice} = 1/3\ C_V v_s d$, where C_V is the heat capacity per unit volume, v_s is the velocity of sound, and d is the mean free path of the heat carrying phonons. If the average atomic weight is high, the heavy atoms lead to small sound velocities and a correspondingly low thermal conductivity [Ashcroft and Mermin, 1976], which makes the material a good thermoelectric. It is also important to consider that mass fluctuation scattering can be used to reduce the lattice thermal conductivity. The idea behind this rule is that iso-valent substitutions will scatter heat carrying phonons strongly because the wavelength of these phonons is about the same as the distance between the scattering centers. Electrons, on the other hand, have a longer wavelength and will be scattered less.

Crystal structures with many atoms per unit cell can tend to have low lattice thermal conductivities. This is not as well-grounded theoretically (or experimentally), but nevertheless seems to be validated by experience. One explanation for this trend is that the number of defects per unit cell tends to grow rapidly as the size of the cell increases. The amount of disorder, then, tends to be relatively greater for materials with many atoms per unit cell.

Another explanation lies in the breakdown of the concept of a phonon as the number of atoms in the unit cell grows large. Remembering that there are 3n phonon modes, where n is the number of atoms in the unit cell, it is reasonable to assume that as n grows large these modes will begin to overlap and will no longer be distinguishable. It has been argued by Allen and Feldman [1993] that thermal transport in this situation is beginning to resemble thermal transport in a glass.

Crystal structures, in which the ions are highly coordinated, tend to have lower thermal conductivities than crystal structures in which the ions have low coordination. This is an empirical relationship proposed by Spitzer [1970] based on a compilation of thermal conductivity data on more than 200 semiconductors. No one is aware of any generally accepted explanation for this behavior, although it is interesting that highly coordinated ions are also involved in the reduction in thermal conductivity associated with the "rattling" cations, which will be discussed below.

A new concept proposed by Slack [1995], involves finding materials in which one or more atoms per unit cell are loosely bound and "rattle" in an oversized cage. The cage is invariably constructed from many atoms that highly coordinate the rattler. Such rattlers resonantly scatter phonons, and can reduce the mean free path of the heat carrying phonons to dimensions comparable to the inter-atomic spacing. The effect on the thermal conductivity is dramatic, as recent work on filled skutterudites [Sales *et al.,* 1996, 2000] and germanium clathrates [Cohn *et al.,* 1999] has shown.

Finally, it is important to recognize that although these characters describe most of the better thermoelectrics, there are some relatively good materials (Na_xCoO_2 and some layered Co oxides) that do not have the above said characters. This material is made up of light atoms with large electro-negativity differences, yet at room temperature, the power factor of this compound is greater than that of a better thermoelectric material Bi_2Te_3 [Terasaki *et al.,* 1997].

1.5.1 CONVENTIONAL THERMOELECTRIC MATERIALS

Thermoelectric materials so far used for practical applications are Bi_2Te_3, PbTe, $Si_{1-x}Ge_x$ and N-type BiSb. Bi_2Te_3 shows the highest performance near room temperature [Yamashita and Sugihara, 2005] and is used for cooling applications such as Peltier coolers which are commercially available. Sb_2Te_3 is also a good thermoelectric material. The dominant defect in Sb_2Te_3 is antimony atoms on tellurium sites. Since an antimony atom has one less valence electron to donate to the crystal, it can be thought of as an acceptor site, trapping a valence band electron and producing a hole. Sb_2Te_3 is therefore normally p-type. Bi_2Te_3 contains both bismuth on tellurium site defects and tellurium on bismuth site defects, which are acceptors and donors respectively, so Bi_2Te_3 can be either p or n-type. A good positive thermoelectric material (i.e., $Q > 0$) is a solid solution of composition 25%:75% Bi_2Te_3:Sb_2Te_3 [Fan *et al.*, 2006]. A good negative thermoelectric material (i.e., $Q < 0$) 25%:75% Bi_2Te_3:Bi_2Se_3, is also a solid solution.

It was proved that the lead telluride (PbTe) is an intermediate thermoelectric power generator material with maximum operating temperature of 900 K. PbTe has a high melting point, good chemical stability, low vapor pressure and good chemical strength in addition to high figure of merit Z. In this class of materials, PbTe alloys play an important role, in high temperature applications. The figure of merit is high in $AgPb_mSbTe_{m+2}$ alloys with ZT=2.2 at 800K [Hsu *et al.*, 2004]. PbTe shows the highest performance near 500-600K and $Si_{1-x}Ge_x$ is superior above 1000 K [Jeffrery Synder and Caillat, 2004].

A material which is a promising candidate to fill the temperature range in the ZT between those based on bismuth telluride and lead telluride is the semiconductor compound β-Zn_4Sb_3 [Chen, 1998]. This material possesses an exceptionally low thermal conductivity and exhibits a maximum ZT of 1.3 at a temperature of 670 K. This material is also relatively inexpensive and stable up to this temperature in vacuum.

The conventional thermoelectric materials are degenerate semiconductors of high mobility. It is suggested that materials containing heavy elements (giving small sound velocity) with their solid solutions (giving short phonon mean free path) and many atoms in a unit cell (giving small lattice specific heat) can be a good candidate. Mahan [1989] has suggested a microscopic parameter for good thermoelectric materials called the B-factor, given as

$$B = \left(\frac{2mk_BT}{\pi\hbar^2}\right)^{\frac{3}{2}} \frac{\mu}{K_{lat}} \tag{1.17}$$

where m, the effective mass, μ, the mobility and K_{lat}, the lattice thermal conductivity are independent parameters. Accordingly, a degenerate semiconductor with heavier effective mass, higher mobility and lower lattice conductivity is extensively searched. Table 1.1

lists the thermoelectric parameters of the conventional thermoelectric materials [Mahan, 1998].

Table 1.1 Thermoelectric parameters of the conventional thermoelectric materials.

Material	Temperature for maximum ZT (K)	Effective mass	Mobility (m^2/Vs)	Lattice thermal conductivity (W/mK)	Figure of merit (ZT)
Bi_2Te_3	300	0.2	0.12	1.5	1.3
PbTe	650	0.05	0.17	1.8	1.1
$Si_{1-x}Ge_x$	1100	1.06	0.01	4.0	1.3

The electrical conductivity (σ), thermo power (S) and thermal conductivity (K) of the conventional thermoelectric materials Bi_2Te_3, PbTe and $Si_{1-x}Ge_x$ are around 1-2 mΩcm, 150-200 μV/K and 15-25 mW/cmK respectively. The B-factor is around 0.3-0.4, which is significantly larger than that of other semiconductors. The thermoelectric property of silicides has been recently found with the exception of the alkali and alkaline earth silicides. Semiconducting behaviour is observed only in the more silicon rich silicides like iron disilicide [Federov and Zsitsev, 2005]. Iron disilicide is a semiconductor at temperatures below 1259 K and a metal above this temperature. The energy gap is around 0.88 eV at room temperature indicating potential high temperature applications.

1.5.2 FILLED SKUTTERUDITE COMPOUND

Since the discovery of Bi_2Te_3 in mid 1950's, thermoelectric materials were extensively searched in binary systems. In fact, many promising materials were found through the research, but ZT did not exceed unity. Filled skutterudite, $Ce_xFe_3CoSb_{12}$ is the first unambiguous example who's ZT exceeds unity and is going to be used for thermoelectric power generation in the next generation [Sales *et al.*, 1997].

The crystal structure of the skutterudite $CoSb_3$ has a unit cell of cubic symmetry and consists of the eight sub cells whose corners are occupied by Co atoms. Six sub cells out of the eight are filled with Sb packets, forming the valence band. According to the band calculation, $CoSb_3$ is a narrow gap semiconductor with an indirect band gap of 0.5 eV, which is favorable for a thermoelectric material. In fact, the hole mobility of $CoSb_3$ exceeds 2000 cm^2/Vs at 300K and Seebeck coefficient is as high as 630 μV/K [Chen *et al.*, 2001] which is much higher than that for Bi_2Te_3 [Caillat *et al.*, 1996].

The crystal structure of the skutterudite $CeFe_3CoSb_{12}$ has two vacant sub cells of the skutterudite and two filled Ce ions. In order to compensate the charge valance, six Fe atoms are substituted for the eight Co sites, because Ce usually exists as trivalent. The most remarkable feature of this compound is that filled Ce ions reduce the lattice thermal conductivity several times lower than that for an unfilled skutterudite $CoSb_3$. Ce ions are weakly bound in an oversized atomic cage so that they will vibrate independently from the other atoms to cause large local vibrations. This vibration and the atom in the cage are named "rattling" and "rattles" respectively. As a result, the phonon mean free path can be as short as the lattice parameters. Namely this compound has a poor thermal conduction like a glass and a good electric conduction like a crystal, which is called "phonon glass electron crystal (PGEC)" named by Slack [1995]. Nevertheless, the concepts of rattling and phonon glass have been a strong driving force for thermoelectric material search in recent years. Accordingly, many promising materials such as $Sr_6Ga_{16}Ge_{30}$ [Nolas *et al.*, 1998] and $CsBi_4Te_5$ [Chung *et al.*, 2000] have been synthesized and identified.

1.5.3 LAYERED CO-OXIDES

As mentioned in the previous section, the state-of-the-art thermoelectric materials Bi_2Te_3, PbTe, and $Si_{1-x}Ge_x$ are degenerate semiconductors of high mobility. Since Te is scarce in earth, toxic, and volatile at high temperature, the application of Bi_2Te_3 and PbTe has been limited. By contrast, oxide is chemically stable at high temperature in air, and thus oxide thermoelectric is expected to be used in a much wider area. However, most of the oxide semiconductors show very low mobility, and have been thought to be out of the question.

The discovery of large thermo power and low resistivity thermoelectric material $NaCo_2O_4$ single crystal [Terasaki *et al.*, 1997], kindle the interest on growing some kinds of oxide thermoelectric materials [Koumoto *et al.*, 2003; Fujita *et al.*, 2001]. The figure of merit (ZT) of $NaCo_2O_4$ single crystal exceeds unity at 800 K. Another fascination of $NaCo_2O_4$ is existence of various related oxides. Following $NaCo_2O_4$, $Ca_2Co_4O_2$ [Funahashi *et al.*, 2000], $(Bi,Pb)_2Sr_2Co_2O_8$ [Funahashi and Matsubara, 2001], $TiSr_2Co_2O_y$ [Hebert *et al.*, 2001], and $(Hg,Pb)Sr_2Co_2O_y$ (Maignan *et al.*, 2002) have been found to show good thermoelectric performance.

The layered Co oxide $NaCo_2O_4$ shows low resistivity when compared to the layered Cu oxide, $Bi_2Sr_2CaCu_2O_8$ (one of high Tc superconductors), whereas the layered Ni and Mn oxides show hopelessly high resistivity. For thermo power, the difference between the Co oxide and the other oxides is more remarkable. $NaCo_2O_4$ shows 100 μV/K at room temperature, while the layered Cu, Ni, and Mn oxides show very small thermo power of the order (1-10) μV/K. Thus, the most peculiar feature of the layered Co oxide is the unusually high thermo power.

Also, rare earth compound $YbAl_3$ [Rowe *et al.*, 2002], although possessing a relatively low figure of merit, has a power factor almost three times that of bismuth telluride while

Mg_2Sn has almost the same performance but costs less than a quarter of the price [Rowe and Min, 1994].

1.6 THERMOELECTRIC APPLICATIONS

1.6.1 RADIO-ISOTOPE THERMOELECTRIC GENERATOR

Space exploration missions require safe, reliable, long-lived power systems to provide electricity and heat to spacecraft and their science instruments. A uniquely capable source of power is the radio isotope thermoelectric generator (RTG), essentially a nuclear battery that reliably converts heat into electricity. The new generation power system developed by National Aeronautics and Space Administration (NASA) called multi-mission radio-isotope thermoelectric generator (MMRTG), is being designed to operate on planetary bodies with atmospheres such as Mars, as well as in the vacuum of space. In addition, the MMRTG is a more flexible thermoelectric modular design capable of meeting the needs of a wider variety of missions as it generates electrical power in smaller instruments, slightly above 100 watts. The design goals for the MMRTG include ensuring a high degree of safety, optimizing power levels over a minimum lifetime of 14 years, and minimizing weight.

The state of art thermoelectric generators have been successfully used by (NASA) to explore the solar system for many years. The Apollo missions (to the Moon), the Viking missions (to Mars), and the Pioneer, Voyager, Ulysses, Galileo, Cassini and Pluto New Horizons (outer solar system) missions all used RTGs. Over the last four decades, the United States has launched 26 missions involving 45 RTGs.

RTGs work by converting heat from the natural decay of radioisotope materials into electricity. RTGs consist of two major elements, one is a heat source that contains plutonium-238 dioxide and the other a set of solid-state thermocouples that convert the plutonium's heat energy to electricity. The first generators Orion-1 and Orion-2 employed a Polonium-210 heat source and powered the onboard equipment to artificial satellites Cosmos-84 and Cosmos-90. The MMRTG contains a total of 4.8 kg (10.6 lb) plutonium-238 dioxide that initially provides approximately 2,000 watts of thermal power and 120 watts of electrical power [Abelson, 2005; Bennett, 2005].

1.6.2 WASTE HEAT RECOVERY SYSTEM

Around 1995, the U.S. Department of Energy (USDOE) initiated a project with Hi-Z technologies to develop a thermoelectric generator (TEG) demonstrator to convert the waste heat from a heavy-duty Class 7-8 diesel engine directly to electricity [Matsubara,

2002]. This unit used Bismuth Telluride cells and provided a nominal 1 kW. This TEG was integrated with the muffler and was installed in a heavy-duty truck. Radiator cooling water (~110°C) was used to extract the heat from the cold side of the TEG. The TEG was run for the equivalent of 500,000 miles on a test track [Bass *et al.,* 1995]. These data coupled with a first approximation analysis justified initiation of a competitive procurement to develop thermoelectric generators (TEG's) for transportation vehicles, to either augment or replace the alternator.

In standard diesel-powered vehicles, up to 65% of the fuel chemical energy used to provide vehicle propulsion is lost as waste heat. Most of this waste heat is rejected in relatively equal proportions through the cooling (radiator) and exhaust systems. If some of this wasted heat can be converted into electrical energy and stored in an auxiliary battery or a hybrid electric vehicle battery pack, the efficiency of the vehicle can be improved.

A typical automotive alternator provides up to 1–2 kW of power for a vehicle's auxiliary equipment. Efficiencies for the alternator, a belt-driven device that runs directly off the vehicle's engine, vary from 60% at low engine speeds down to 30% at higher engine speeds. Coupled with the low efficiency of the engine itself, the overall efficiency of the energy producing process of the alternator may be as low as 10-15%. Thus, to produce 1 kW of power from the alternator may require as much as 10 kW of fuel chemical energy. If a waste heat recovery device could supply enough power to support the auxiliary power requirements of the conventional vehicle, then the alternator could be removed from the vehicle with the associated improvements in overall vehicle efficiency. In addition, replacing a maintenance-intensive alternator with a solid-state and rugged waste heat recovery device would also be a significant benefit for the user.

Thermoelectrics present possibilities of recovering electricity from energy intensive industrial processes. Converting waste heat to electricity in the aluminium process should reduce the cost of aluminium such that it could be considered for mass market vehicles. Currently Jaguar, Aston Martin and Audi A-8 have aluminium frames and bodies which reduce vehicle weight by about 500 pounds. Oak Ridge National Laboratory has empirically developed a rule of thumb that a 10 percent reduction in vehicle weight improves fuel economy by 5 to 7 percent.

1.6.3 MOBILE AIR CONDITIONERS

There is also concern about the global warming contribution of conventional mobile air conditioners. Cooling/heating using currently available thermoelectric materials could provide significant advantages compared with current systems for improved fuel economy, reduced toxic and greenhouse gas emissions. Thermoelectric cooling systems could be designed to take best advantages of thermoelectrics. Compact thermoelectric units can be installed in the seats, dashboard and overhead for the driver and the front seat occupant. Units can be installed in the back of the front seats, the overhead, seats and floor. These units can be devised to only cool or heat the person, not the whole cabin. The driver of a car can be cooled with less than 700 watts of cooling whereas current air conditioners utilizes up to 3,500 to 4,000 watts. Thermoelectric cooling systems can be converted from air conditioning to heating by simply changing the polarity of the DC power. The thermoelectric modules are silent with no moving parts. However, the thermoelectric cooling system would have fans and a coolant loop circulating pump to remove the heat from the modules and to a heat exchanger to disburse the heat.

1.6.4 OTHER COMMON APPLICATIONS

High figure of merit (ZT) thermoelectrics could challenge photovoltaics for a wide range of household and commercial applications. Thin film high efficiency thermoelectrics could bond to the back of solar cells and either air or liquid cooling could be applied to the thermoelectric cold side. Solar cell efficiency is typically an inverse function of temperature so the thermoelectric thin film converting conducted heat to electricity would be concurrently reducing the solar cell temperature. Low power applications include a thermoelectric watch battery [Saiki *et al.,* 1985] utilizes waste human body heat with power less than a watt. The battery's thermo-couples are prepared by depositing germanium and indium antimonide on either side of a 1mm thick insulator, which served as a stimulated watch strap. The required 2 volt power supply is obtained by connecting 2875 thermo-elements in series.

Many homes use solar energy to heat water in rooftop coils. This activity could be enhanced with solar concentrators. Flat plate thermoelectric modules could be installed such that the exiting hot water provides the T_{hot} and the cooled water would be routed through the TEG back to the solar heater. In colder climates the solar heating coils could be encapsulated in a greenhouse effect structure to optimize solar energy and minimize convective heat transfer losses. Thermoelectrics may also become the most effective approach for geothermal energy conversion to electricity.

1.7 X-RAY POWDER DIFFRACTION (XRD)

1.7.1 X- RAYS

X-ray powder diffraction is a powerful nondestructive testing method for determining a range of physical, chemical and structural characteristics of the material. Monochromatic X-rays are used to determine the inter planer spacing of the unknown materials.

Crystalline samples are analyzed either in a single crystal form with perfect periodicity in three dimension or as powder with grains in random orientations to insure that all crystallographic directions are sampled by the beam.

It is widely used in all fields of today's science and technology. The X-ray spectra generated by this technique provide a structural fingerprint of the unknown mixture of crystalline materials and the relative peak height of multiple materials may be used to obtain semi-quantitative estimates of abundances. Data reduction routines rapidly determine peak position, relative intensities and intra crystalline d-spacing. The applications include phase analysis, i.e., the type and quantities of phases present in the sample, the crystallographic unit cell, crystal structure, crystallographic texture, crystalline size, macro stress, micro strain and also electron radial distribution function. The major advantages of X-ray diffraction are rapid identification of materials, ease of sample preparation, computer-aided material identification, large library of known crystalline structures and multi-sample stage.

1.7.2 X-RAY DIFFRACTION

X-ray diffraction results from the interaction between X-rays and electrons of the atoms. W.L. Bragg and W.H. Bragg studied the diffraction of X-rays in detail and used a crystal of rock salt to diffract X-rays and succeeded in measuring the wavelength of X-rays. It was found that the X-rays when passing through the crystals undergo diffraction, similar to the one that happens with the grating, and the diffracted rays interfere constructively to give measurable intensities. Also, it was found that the constructive interference takes place only when the lattice system obeys the condition called Bragg's law [Bragg, 1913].

Depending upon the atomic arrangement, the scattered rays interfere constructively, when the Bragg's condition is satisfied, *ie.,* the path difference between two diffracted rays differ by $2d\sin\theta$, which is an integral multiple of its wavelength λ. When the Bragg conditions for constructive interference are obtained, a reflection is produced with the relative peak height proportional to the number of grains in a preferred orientation. Figure 1.1 represents the X-ray diffraction from the two nearby lattice planes which are separated by Bragg spacing.

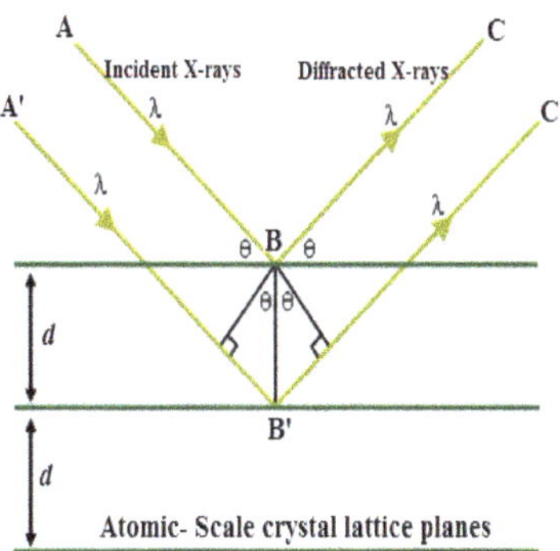

Figure 1.1 X-ray diffraction from lattice planes.

The selective condition is described by the Bragg equation, also called Bragg's law,

$$2d_{hkl}\sin\theta_{hkl} = n\lambda \qquad (1.18)$$

Where λ is the wavelength of the incident X-rays, d_{hkl}, the Bragg spacing and θ_{hkl} the Bragg angle, which is half the angle between incident and reflected beams. The hkl represents the miller indices triplet of the lattice planes [Stout and Jensen, 1970].

1.7.3 POWDER DIFFRACTION

An ideal powder for a diffraction experiment consists of a large number of small randomly oriented crystallites (coherently diffracting crystalline domains). If the domains are sufficiently large, there are always enough crystallites in any diffracting orientation to give reproducible diffraction patterns. To obtain a precise measurement of intensity of diffracted rays, the crystallite size must be small, typically of the order of 10 μm or less and depending on the characteristics of the specimen like absorption, shape, etc., and the diffracting geometry.

Another important criteria is that the pattern obtained by step scattering of the incident monochromatic X- rays should be detected with small increments of $\Delta(2\theta)$. The increments, i.e., the step size may be between 0.05° to 0.001° in 2θ.

A characteristic feature of powder diffraction is the collapse of the three dimensional reciprocal space of individual crystallites on the one dimensional 2θ axis. The resulting effects are:

1. Systematic overlapping of the diffracted peaks due to symmetry conditions, as a cubic space group.
2. Overlapping due to limited experimental resolution.
3. Considerable background intensity which is difficult to define with greater accuracy.
4. Non random distribution of the crystallites in the specimen, generally known as preferred orientation.

1.7.4 STRUCTURE FACTOR AND ELECTRON DENSITY CALCULATION

The structure factor may be defined as the sum of the wavelets scattered from all the infinitesimal elements of electron density in the unit cell. The structure factor, F_{hkl}, is the resultant of j waves scattered in the direction of the reflection hkl by the j atoms in the unit cell. Each of these waves has amplitude proportional to f_j, the scattering factor of the atom.

The exponential form of structure factor is given by,

$$F_{hkl}=\Sigma f_j \exp\,[2(hx_j+ky_j+lz_j)] \tag{1.19}$$

Where f_j denotes scattering factor of jth atom in the unit cell and the structure factor, F_{hkl} is the resultant of the addition of scattering factor associated with each jth atom in the unit cell. Instead of treating the structure factor as said above, it is also possible to consider it as the sum of the wavelets scattered from all the infinitesimal elements of electron density in a unit cell. So, the structure factor can be written as

$$F_{hkl} = \int_v \rho(X,Y,Z)\exp\left[2\pi i\left(hx_j + ky_j + lz_j\right)\right]dv \tag{1.20}$$

where $\rho(X,Y,Z)$ represents the number of electrons per unit volume.

The structure factor can be calculated from electron density and vice versa i.e. from the given set of structure factors, (F_o). Because of the periodicity of crystal structure, the above task can be performed using Fourier series/Fourier transform. With the help of Fourier transform, the electron density $\rho(X,Y,Z)$ can be expressed in terms of structure factor as below.

$$\rho(X,Y,Z) = \frac{1}{V}\Sigma\Sigma\Sigma F_{hkl}\left[-2\pi i\left(hx+ky+lz\right)\right] \tag{1.21}$$

Here F_{hkl} in this equation is not a modulus value, whereas the observed structure factors are modulus ones. In order to avoid this discrepancy, one can write the structure factor as

$$F_{hkl} = |F_{hkl}| \exp(i\alpha) \quad (1.22)$$

The three dimensional electron density equation for the two dimensional calculation reduces as below.

$$\rho(X, Y) = \frac{1}{A} \Sigma \Sigma F_{hk0} \exp[-2\pi i(hx + ky)] \quad (1.23)$$

1.7.5 FOURIER ANALYSIS OF ELECTRON DENSITY DISTRIBUTION

The electron density is defined as the number of electrons per unit volume, which gives more information about the chemical bonding as well as the physical and chemical properties of molecular/solid state system. It can be determined experimentally from elaborate X-ray diffraction measurements at short wavelengths and low temperature. In some cases it can also be calculated from theory. Due to its fundamental importance, the electron density is used across many disciplines in physics, chemistry, geology and also biology [Stout and Jenson, 1970].

The magnitude of individual structure factors is calculated as the square root of the measured diffraction intensities and their phases are determined by solving the structure. The interpretation is described as a model, which is improved by least squares refinements, based on the observed structure factors. The electron density can then be calculated as a Fourier summation of phased structure factors [Stout and Jenson, 1970].

The electron density, a periodic function of the number of electrons in any volume element is

$$\rho\,(X, Y, Z)\, dv$$

The exponential form of the wavelet scattered by this element is

$$\rho\,(X, Y, Z) \exp(-2\pi i(hx+ky+lz))$$

The resultant is the sum of all the elements in the unit cell i.e., the integral over its volume.

$$F_{hkl} = \int_{v} \rho(X, Y, Z) \exp(-2\pi i(hx+ky+lz)) \quad (1.24)$$

1.7.6 LEAST-SQUARES REFINEMENT

Let us imagine an error-free set of $|F_o|$ and an almost correct set of atomic co-ordinates and temperature factors. For simplicity of our discussion, one can assume that the structure is centro-symmetric and the temperature factors are isotropic. The calculated structure factors will be given by

$$(F_c)_{hkl} = \sum_{j=1}^{N/2} 2f_j \left(-B_j \frac{\sin^2\theta}{\lambda^2}\right) \cos 2\pi(h\vec{x}_j + k\vec{y}_j + l\vec{z}_j) \quad (1.25)$$

The correct thermal and positional parameters *i.e,* B_j, $\vec{x}_j$, $\vec{y}_j$ and $\vec{z}_j$ for the jth atom can be expressed as $B_j + \Delta B_j$, $\vec{x}_j + \Delta\vec{x}_j$, $\vec{y}_j + \Delta\vec{y}_j$ and $\vec{z}_j + \Delta\vec{z}_j$ and hence one can write equation (1.25) as

$$(F_o)_{hkl} = \sum_{j=1}^{N/2} 2f_j \left((B_j + \Delta B_j)\frac{\sin^2\theta}{\lambda^2}\right) \cos 2\pi\left[h(\bar{x}_j + \Delta\bar{x}_j) + k(\bar{y}_j + \Delta\bar{y}_j) + l(\bar{z}_j + \Delta\bar{z}_j)\right] \quad (1.26)$$

In the above equation (1.26), N equivalent reflections of a given Laue class are separated into two subsets of $N/2$ equivalent reflections. And now,

$$\Delta F_{hkl} = \sum_{j=1}^{N/2} \frac{\partial(F_c)_{hkl}}{\partial B_j}\Delta B_j + \frac{\partial(F_c)_{hkl}}{\partial \bar{x}_j}\Delta\vec{x}_j + \frac{\partial(F_c)_{hkl}}{\partial \bar{y}_j}\Delta\vec{y}_j + \frac{\partial(F_c)_{hkl}}{\partial \bar{z}_j}\Delta\vec{z}_j \quad (1.27)$$

$$\text{where, } (F_o)_{hkl} - (F_c)_{hkl} = \Delta F_{hkl} \quad (1.28)$$

Equations of type (1.27) and (1.28) can be produced for each reflection and in general, there will be many more equations than parameters. If a subset of these equations was taken such that the number of equations equaled the number of parameters, then the correct parameters could be found and recalculated. Values of F_c's would then equal to the $|F_o|$'s. However in practice, the $|F_o|$'s are not error free and so one finds a least–square solutions of the complete set of equations. This solutions is the one such that when the parameters are changed to $B_j + \Delta B_j$, $\vec{x}_j + \Delta\vec{x}_j$, $\vec{y}_j + \Delta\vec{y}_j$ and $\vec{z}_j + \Delta\vec{z}_j$ the quantity,

$$R_S = \sum_{hkl}\left[(F_o)_{hkl} - (F'_c)_{hkl}\right]^2 \quad (1.29)$$

is a minimum, where $(F'_c)_{hkl}$ is the revised calculated structure factor. From Taylor's theorem, one can find, ignoring second order small quantities, then

$$(F'_c)_{hkl} = (F_c)_{hkl} - \sum_{j=1}^{N/2} \frac{\partial(F_c)_{hkl}}{\partial B_j}\Delta B_j + \frac{\partial(F_c)_{hkl}}{\partial \bar{x}_j}\Delta\bar{x}_j + \frac{\partial(F_c)_{hkl}}{\partial \bar{y}_j}\Delta\bar{y}_j + \frac{\partial(F_c)_{hkl}}{\partial \bar{z}_j}\Delta\bar{z}_j \quad (1.30)$$

Hence,

$$(F_o)_{hkl} = (F'_c)_{hkl} - \sum_{j=1}^{N/2} \frac{\partial(F_c)_{hkl}}{\partial B_j}\Delta B_j + \frac{\partial(F_c)_{hkl}}{\partial \bar{x}_j}\Delta\bar{x}_j + \frac{\partial(F_c)_{hkl}}{\partial \bar{y}_j}\Delta\bar{y}_j + \frac{\partial(F_c)_{hkl}}{\partial \bar{z}_j}\Delta\bar{z}_j \quad .(1.31)$$

Here in the above equation (1.29), R_S is the sum of the squares of the difference between the left-hand sides and right-hand sides of equations (1.30) and (1.31), and it is the minimization of this quantity which is usually implied by the least squares solution.

It is possible to weigh the equations according to the expected reliability of the quantity ΔF_{hkl}. If the measured $|F_o|$ is expected to have a large random error, then the value of ΔF_{hkl} may well be dominated by this and ΔF_{hkl} may not be very useful in indicating the changes to be made in the atomic parameters. Errors of measurement tend to be related to the actual value of F_o and many weighing schemes have been proposed based on the value of $|F_o|$. When a reasonable set of weights has been found, the equation for R_s is multiplied by the appropriate weight and a least squares solution of the modified set of equations is then sought in the usual way. The standard least squares approximate procedure by the full matrix method is adopted for refining the parameters such as scale factors, thermal parameters and extinction parameter. In the least square procedures, the quantity to be minimized is,

$$R_{SW} = \sum_{hkl} W_{hkl} (| F_o| - | kF_c |)^2 \qquad (1.32)$$

Where in the above equation (1.32), W_{hkl} is the weight to be assigned on an observation, F_o is the observed structure factor, F_c is the calculated structure factor and k is the scale factor. From the final value of $R_{SW,}$ the standard errors of the final parameters can be estimated.

Equation (1.32) is a measure of the degree to which the distribution of difference between $|F_o|$ and $|F_c|$ fits the distribution expected from the weights used in the refinement [International Tables for X-ray Crystallography, 1974].

1.7.7 LIMITATION OF FOURIER METHOD

In accurate valence electron distribution, the negative electron density region as a result of termination of the Fourier sum is the major disadvantage of Fourier synthesis. Moreover, weak electron peaks cannot be visualized using the Fourier synthesis. Another drawback is the usage of limited number of Fourier co-efficient and ignoring experimental errors by setting all the structure factors to zero simply because the experiment cannot be carried out. This is a biased assumption.

1.7.8 ELECTRON DENSITY DISTRIBUTION

Since the lattice has a periodicity, the electron density is also considered to behave as a periodic function. The number of electrons in any volume element dV is $\rho(x,y,z)\, dV$. In an X-ray scattering experiment, the wavelet scattered by this element is

$$\rho\,(x,y,z) \exp[2\pi i(hx + ky + lz)]dV \qquad (1.33)$$

The resultant sum of contributions from all the elements in the unit cell, i.e., the integral over its volume gives

$$F_{hkl} = \int \rho(x, y, z) \exp[-2\pi i(hx + ky + lz)]dV \qquad (1.34)$$

The structure factor is considered as a resultant of adding the scattered waves in the direction of the *hkl* reflection from the atoms in the unit cell. This approach was based on the assumption that the scattering power of the electron cloud surrounding each atom could be equated to that of the proper number of electrons concentrated at the atomic center. But the structure factor may equally well be considered as the sum of the wavelets scattered from all the infinitesimal elements of electron density in a unit cell, with no assumptions being made about the distribution of this density. The electron density $\rho(r)$ is defined as the number of electrons per unit volume.

The geometric properties of unit cells can be deduced from the locations of reflections on various kinds of X-ray diffraction photographs. It is concerned with the measurement of the relative intensities of these reflections, since it is from these intensities, the electron density distribution in the crystal cell will be deduced. There are connections between the intensities and the electron density distribution. There are a few general precautions, which are applicable to any intensity measuring method. If the structure factors and phases are known, the electron-density distribution of the unit cell can be calculated. There is necessarily a one-to-one relationship between structure magnitudes and electron density, *i.e.*, a given set of magnitudes must correspond to one and only one electron distribution.

The magnitude of individual structure factors are calculated as the square-root of the measured diffraction intensity and their phases are determined by solving the structure. The interpretation is described as a model, which is improved by least-squares refinement based on the structure factors. The electron density can then be calculated as a Fourier summation of phased structure factors.

Intensities of diffracted X-rays are due to interference effects of X-rays scattered by all the different atoms in the structure. The diffraction pattern is the Fourier transform of the crystal structure, corresponding to the pattern of waves scattered from an incident X-ray beam by a single crystal. It can be measured by experiment (only partially, because the amplitudes are obtainable from the directly measured intensities via a number of correction, but the relative phases of the scattered waves are lost), and it can be calculated (giving both amplitudes and phases) for a known structure. In turn, the crystal structure is the Fourier transform of the diffraction pattern and is expressed in terms of electron density distribution concentrated in atoms, it cannot be measured by direct experiment, because the scattered X-rays cannot be refracted by lenses to form an image as done with light in an optical microscope, and it cannot be obtained directly by calculation, because the required relative phases of the waves are unknown. One can calculate the electron density distribution for a given set of structure factors, using the Fourier series.

1.7.9 RIETVELD ANALYSIS

The explosion of interest in powder diffraction methods during the last 30 years has been driven by a number of factors. The major one was most certainly the development of the Rietveld method [Rietveld, 1969] in the late 1960's, since at a stroke, this extended the scope of powder techniques from simple, high symmetry materials to compounds of substantial complexity in any space group. Within five years, for example, the method was being used to refine the structures of orthorhombic and monoclinic materials with as many as 22 atoms in the asymmetric unit [Von Dreele and Gheetham, 1974], and by 1977, Gheetham and Taylor were able to review the applications of the Rietveld method to over 150 compounds [Gheetham and Taylor, 1977]. The majority of these early applications involved the use of neutrons, but the field received a further boost in the late 1970's and early 1980's with the extension of the Rietveld method to X-ray data [Malmros and Thomas, 1977: [Young *et al.*, 1977], time of flight neutron data [Von Dreele *et al.,* 1982], and then Synchrotron X-ray data [Cox *et al.,* 1983]. These instrumental advances were accompanied by software developments.

The Rietveld method [Rietveld, 1969] for structural refinement of powder diffraction data has been developed over the last four decades and has proved indispensable in solving crystal structures. The process involves minimizing the difference between a crystallographic model and experimental data, via a least squares refinement using the intensity recorded at several thousand equal increments of scattering angle 2θ with a defined model having symmetry, atomic positions, unit-cell size and site occupancies.

It is a structural refinement procedure which uses step intensity data y(i), whereby each data point is treated as an observation. The idea behind the Rietveld method is to consider the entire powder diffraction pattern using a variety of refinable parameters. That way the intrinsic problem of any powder diffraction pattern with its systematic and accidental peak overlaps is overcome. It is the intention to extract as much information as possible from a powder pattern. The program employs directly the individual intensities y(i) at each 2θ value obtained from step scanning measurements of powder diffraction patterns.

Diffraction of a polycrystalline sample, with X-rays, reduces the three dimensional reciprocal lattice to a one dimensional diagram. As a consequence such patterns suffer from overlapping peaks, sometimes accidental due to a lack of resolution, sometimes intrinsic in patterns for samples with cubic or trigonal symmetry. Rietveld's [Rietveld,

1969] approach is based on the complete diffraction pattern, including background. Later a two stage method was extended by adding as a first step, profile fitting resulting in a complete separation of the diffraction peaks. This step yields reliable integrated intensities comparable to single crystal data, and this makes this method more general and closer to single crystal analysis. The method is consequently called today "two stage method".

There is much more information hidden in a powder pattern than just atomic positions, site occupancies and Debye-Waller factors. To name a few: lattice parameters and space group can be deduced from the refined peak positions of the reflections; the amorphous fraction in the specimen or local order/ disorder can be deduced from the background; particle size, strain/stress and domain size in the sample from analyzing the broadening of the peaks, FWHM, and the qualitative and quantitative phase analysis. To understand the Rietveld method [Rietveld, 1969], several inherent properties and problems have to be discussed as:

1. Peak shape
2. Peak width(FWHM)
3. Preferred orientation
4. Method of calculation

The measured profile of a single, as well powder diffraction peak is dependent on two intrinsic parameters (i) an instrumental parameter including the spectral distribution, i.e., the monochromatic mosaic distribution, and the transmission function determined by the slits, and (ii) the sample contribution based on the crystal structure and crystallinity of the sample. While these contributions can have a form not necessarily being Gaussian, it is an empirical fact that their convolution produced in neutron diffraction patterns almost exactly resemble a Gaussian peak shape. This is different in X-ray diffraction, where especially the instrumental contributions lead to rather complicated peak profiles. A number of profiles have been suggested and tested in the past and some are still preferred by most scientists. The majority, however just uses the Pseudo-Voigt function, the summation of a Lorentzian and Gaussian function with adjustable contributions.

The width of the diffraction peaks is the second important parameter and variable when describing a diffraction pattern. The peak width, described as full width half maximum is in general a function of the diffraction angle 2θ. Caglioti *et al.*, [1958] has studied the angle dependence of FWHM for neutron diffraction. They have given a formula to describe this angular dependence.

$$(FWHM)_k = U\tan^2\theta_k + V\tan\theta_k + W \quad (1.35)$$

with U, V and W being adjustable parameters. This simple formula describes adequately the experimentally observed variation of half width with scattering angle. The parameters are refinable quantities in Rietveld's [Rietveld, 1969] least squares calculations. The initial and approximate starting values are found at the beginning of the experimental cycle by measuring FWHM in a standard sample with individual single peaks. These values are unchanged as long as the experimental setup is unchanged and there is no line broadening coming from the crystallites.

In powder samples, there is a tendency for plate or rod like crystallites to align themselves along the axis of a cylindrical sample holder. In solid polycrystalline samples the production of the material may result in greater volume fraction of certain crystal orientations. In such cases, the reflection intensities will vary from that predicted for a completely random distribution. Rietveld [Rietveld, 1969] allowed for moderate cases of the former by introducing a correction factor,

$$I_{cor} = I_{obs} \exp(-G\alpha^2) \quad (1.36)$$

where I_{obs} is the intensity expected for a random sample, G is the preferred orientation parameter and α is the acute angle between the scattering vector and the normal of the crystallites.

For Rietveld [Rietveld, 1969] refinements, the data must be in digital form. The basis for the refinement are the numerical intensity y_i at each of the several thousand equal steps along the scattering angle 2θ, with increments Δ2θ. Typical step sizes range from 0.01° to 0.05°. The best fit sought is the best least squares fit to all of the thousands of yi simultaneously. The quantity minimized therefore is in general terms

$$S=\sum w_i[y_i(obs)-y_i(cal)]^2 = \text{minimum} \quad (1.37)$$

with w_i being the weight of each observation point, $y_i(obs)$ and $y_i(cal)$ are the observed and from a model calculated intensities at each step. The sum i is over all data points. The main parameters in today's programs can be divided into three groups: The first group defines basic experimental parameters: the profile parameters, the half width parameters (U, V, W), possible asymmetries (P) of the diffraction peaks and a zero point adjustment (Z). The second group contains the unit cell parameters (a, b, c, α, β, γ), the crystallographic symmetry, especially space group and preferred orientation parameter (G).

Finally, the third group contains the actual structural parameters like the overall scale factor c, with y(cal)=cy(obs), fractional positional coordinates of the jth atom in the asymmetric unit (x_j, y_j, z_j), atomic (isotropic) Debye-Waller factors (anisotropic parameters have also been included), B_j and occupation number of each crystallographic site (Nj) with j^{th} atom in the asymmetric unit. In Rietveld [Rietveld, 1969] program, the approximate values of all the parameters are required for the first refinement cycle. In subsequent refinement cycles these parameters are refined until a certain convergence criterion is reached, or the refinement is stopped by the operator.

Background is a crucial point in the refinement. Background becomes more and more important since more and more specimens have high background or amorphous material, whose scattering shows up in the background. The background parameters of the profile are fitted to suitable polynomials being B_0, B_1, B_2, B_3, B_4, B_5 and B_6 in six parameter model. Six parameter models as well as 12 parameter models etc. are available in a Rietveld [Rietveld, 1969] profile fitting methodology. The cell parameters and other structural parameters can also be refined using this method. The peak width can be adjusted and refined using the peak shape parameters u, v, w etc. The asymmetric peak widths can be refined using Bearer-Baldinozzi (asym1, asym2, asym3 and asym4) and the profile fitting model such as voight, pseudo-voight etc., can also be used for profile fitting. The micro absorption effects, surface roughness, zero shift, scale factor, temperature factors, occupancy of atoms and composition of atoms are the examples of some other parameters that can be refined.

In the present work, the powder x-ray intensity data of Mg_2Si, PbTe, $Bi_{1-x}Sb_x$, Sb_2Te_3 and Bi_2Te_3 and single crystal data of InSb and $Sn_{1-x}Ge_xTe$ obtained from X-ray diffraction were corrected for Lorentz-polarization and multiplicity factor (m). Other corrections have been done while refining the structure. The structures were refined using JANA 2006 [Petříček *et al.,* 2006] that can refine several structural parameters, the scale, thermal parameters, *etc.*

JANA2006 [Petříček et al., 2006] can handle multiphase structures (for both powder and single crystal data), as well as twins with partial overlap of diffraction spots, commensurate and composite structures. It contains powerful transformation tools for symmetry (group-subgroup relations), cell parameters and commensurate- supercell relations. Wide scale of constrains and restrains is available including a powerful rigid body approach and possibility to define a local symmetry affecting only part of the structure. The latest development of Jana2006 [Petříček et al., 2006] also concerns magnetic structures.

The results of these refinements are given in appropriate sections, and then the refined structure factors have been utilized for maximum entropy method (MEM) refinements to elucidate the electron density distribution, which will be discussed in the next section.

1.8 MAXIMUM ENTROPY METHOD (MEM)

The maximum entropy method (MEM) plays a vital role in determining the precise electron density distribution inside a unit cell both qualitatively and quantitatively. The Maximum entropy method (MEM) is a method to derive the most probable map based on a non-linear calculation by the use of information and probability theory for a given set of experimental diffraction data. It is originally designed to reconstruct the most probable and least biased probability distribution in an underdetermined situation. MEM plays a vital role in crystallography in determining the electron density in the unit cell, which are strictly positive, that provide the best fit to the single crystal or powder diffraction data. It is a model-independent approach in contrast to structure refinements, in which the positions of spherical atoms are determined.

Many reflections usually overlap in the powder diagram and since not all phases can be determined, the dataset used to calculate an electron density is systematically incomplete, and much information is lost compared with the single crystal case. A valuable feature of the MEM is, that all remaining information obtained from such a measurement can be used in the calculation of the charge density, i.e., phased, un-phased, ambiguously phased and overlapping reflections, while in a Fourier calculation only the phased reflections can be used, which leads often to severe distortions of the calculated Fourier-density.

Therefore the MEM is most suited and this method has been used in many fields of crystallography, as reviewed by Gilmore [1996] to study the aspects of structures that go beyond the independent atom approximation, like partially ordered structures, the electron density in the chemical bonding region, and the effects of harmonic and anharmonic atomic vibrations, as well as the study of modulation functions in aperiodic crystals.

Usually, a MEM-charge density is less affected by missing reflections and series termination errors as compared with an ordinary Fourier density. Some interesting crystallographic applications published are, thermal vibrations from single crystal neutron diffraction data [Takata *et al.,* 1994], model-free search for extra-framework cations in zeolites using powder diffraction [Papoular and Cox, 1995], extraction of strictly positive integrated intensities from strongly overlapping reflections in powder patterns [Sivia and David, 1994], disorder in crystals [Papoular *et al.,* 1997], the problem of overlapping reflections in Laue diffraction patterns [Bourenkov *et al.*, 1996] and magnetization densities from polarized neutron diffraction data [Schleger, *et al.*, 1997].

1.8.1 ALGORITHM OF MEM

The algorithm used in our studies has been described by Sakata and Toraya [1990] and Sakata, *et al.*, [1990]. Its single-pixel-approximation has been further discussed by Kumazawa, *et al.*, [1995]. The basis is an iterative algorithm of Collins [Collins, 1982]. The unit cell of a crystal is divided in pixels of equal size in x, y and z directions of the unit cell, and the positive charge density is used in a normalized form. This discrete density is then used as if it were a probability distribution of independent events, so that the Maximum-Entropy principle of Jaynes can be applied to it [Skilling, 1989].

The most probable distribution fitting the known data is the one with the maximum value of the entropy

$$S=-\sum \rho'(r)\ln\frac{\rho'(r)}{\tau'(r)}$$

with $\rho'(r)$ being the normalized pixel density value, and $\tau'(r)$ a density incorporating prior knowledge. The $\rho'(r)$and $\tau'(r)$are normalized as

$$\sum \rho'(r)=1, \qquad \sum \tau'(r)=1.$$

Let the probability $\rho'(r)$and prior probability $\tau'(r)$ be related to the actual electron density in a unit cell as,

$$\rho'(r)=\frac{\rho(r)}{\sum_r \rho(r)} \quad \text{and} \quad \tau'(r)=\frac{\tau(r)}{\sum_r \tau(r)}$$

where $\rho(r)$ and $\tau(r)$are the electron and prior electron densities at a certain pixel (r) in a unit cell, respectively. In the present theory, the actual densities are treated instead of the normalized densities. The soft constraint is introduced as

$$C=\frac{1}{N}\frac{\sum |F_{cal}(k)-F_{obs}(k)|^2}{\sigma^2(F_{obs}(k))} \tag{1.38}$$

Where N is the number of reflections, σ (k) is the standard deviation of F_{obs} (k), F_{obs} (k) is the observed structure factor, F_{cal} (k) the calculated structure factor given by

$$F_{cal}(k)=V\sum \rho(r)\exp(2\pi i\,\mathrm{k.r})\mathrm{dv} \tag{1.39}$$

where V is the volume of the unit cell.

This type of constraint is sometimes called a weak constraint, in which the calculated structure factors agree with the observed ones as a whole when C becomes unity.

As can be seen in equation (1.38), the structure factors are given by the Fourier transform of the electron density distribution in the unit cell. Equation (1.38) guarantees that it is possible to allow any kind of deformation of the electron densities in real space as long as information concerning such a deformation is included in the observed data. One can use Lagrange's method of undetermined multiplier (λ) in order to constrain the function C to be unity while maximizing the entropy.

For this, one can have

$$Q = S - \left(\frac{\lambda}{2}\right) C$$

$$= -\sum \rho'(r) \ln\left(\frac{\rho'(r)}{\tau'(r)}\right) - \frac{\lambda}{2N} \sum_k \frac{\left|F_{cal}(k) - F_{obs}(k)\right|^2}{\sigma^2(k)} \qquad (1.40)$$

and $\frac{dQ}{d\rho(r)} = 0$ yields

$$-\sum_r \frac{\rho(r)}{\sum \rho(r)} \frac{\tau(r)/\sum \tau(r)}{\rho(r)/\sum \rho(r)} \frac{1}{\tau(r)} - \sum_r \frac{1}{\sum \rho(r)} \ln \frac{\rho(r)}{\tau(r)} \qquad (1.41)$$

$$-\frac{1}{\sum \rho(r)}\left[1 + \ln \frac{\rho(r)}{\tau(r)}\right] - \frac{\lambda V}{N} \sum \frac{\left|F_{cal}(k) - F_{obs}(k) e^{-2\pi i \vec{k}.\vec{r}}\right|}{\sigma^2(k)} = 0$$

Using the approximation $\ln x \approx x - 1$

$$\frac{-\rho(r)}{\tau(r)} = \frac{\lambda V z}{N} \sum \frac{\left|F_{cal}(k) - F_{obs}(k) e^{-2\pi i \vec{k}.\vec{r}}\right|}{\sigma^2(k)}$$

$$\rho'(r) = \tau'(r)\left(\frac{\lambda F_{000}}{N}\right) \sum \frac{1}{\sigma^2(k)} \left|F_{cal}(k) - F_{obs}(k) e^{-2\pi i \vec{k}.\vec{r}}\right| \qquad (1.42)$$

where F_{000} is the total number of electrons in the unit cell. Equation (1.40) cannot be solved as it is, since $F_{obs}(k)$ is defined on $\rho(r)$.

In order to solve equation (1.40) in a simple manner, we introduce the following approximation

$$F_{cal}(k) = V \sum_r \tau(r) \exp(-2\pi i k.r). \qquad (1.43)$$

This approximation can be called the zeroth order single pixel approximation (ZSPA). By using this approximation, the right hand side of equation (1.40) becomes independent of $\tau(r)$ and equation (1.40) can be solved in an iterative way starting from a given initial density for prior distribution. For the initial density of the prior density, $\tau(r)$ a uniform density distribution is employed in this work. The choice of prior distribution corresponds to the maximum entropy state among all possible density distributions. It is obvious that there is no prejudice for the initial density.

1.8.2 PRACTICAL ASPECTS OF MEM

The programs read a MEM-input file with the crystal parameters, reflection list with real and imaginary part of the structure factor, an asymmetric unit density file and a documentation file. The MEM refinements were carried out by dividing the unit cell into 128x128x128 pixels. The initial electron density at each pixel is fixed uniformly as F_{000}/a_0^3. Where F_{000} is the total number of electrons in the unit cell and a_0 is the cell parameter. The asymmetric unit pixel coordinates are determined from a general algorithm by testing each pixel in the whole unit cell. The space group is input *via* its number and setting number as listed in the International Tables for Crystallography [1993]. The Lagrange parameter is suitably chosen so that the convergence criterion C=1 is reached after minimum number of iterations. If the parameter λ in equation (1.40) is chosen too small, the calculation converges too slowly, if it is too large, the constraint sum C changes too quickly, or produces a divergence error, when C gets larger instead of smaller. The calculation of this cycle is then automatically repeated with λ set to half its value, until the proper convergence is achieved again. If the last cycle was successful, then λ is automatically increased by a small amount of 1.5 %. With this procedure the program can run in a batch queue without inspection by the user, avoiding both too small and too large values of λ.

The numerical MEM computations are carried out using the software package PRIMA [Izumi and Dilanian, 2002] and the three dimensional (3D) and two dimensional (2D) representation of the electron densities are pictorially visualized using the program VESTA [Momma and Izumi, 2006].

1.9 PAIR DISTRIBUTION FUNCTION (PDF)

The technique of choice for studying local structure of materials has been X-ray absorption fine structure experiments (XAFS) [Koeningsberger and Prins, 1988]. However, XAFS provides information about the immediate atomic ordering (first and sometimes second coordination shells) and all longer- ranged structural features remain hidden. To remedy this shortcoming, one can utilize the alternative approach of obtaining pair distribution function (PDF) from X-ray diffraction data. Determining the PDF has been the approach of choice for characterizing glasses, liquids and amorphous materials for a long time [Debay and Menki, 1930; Bowran and Finne, 2003]. However, its wide spread application to crystalline materials, where some deviation from the average structure is expected to take place, has been relatively recent [Proffen and Billinge, 1992; Petkov *et al.*, 2002]. This real space method utilizes a very small number of experimental techniques that can be used to probe structure on the nanometer length scale, when the local structure is not consistent with the long range, globally averaged structure [Egami and Billinge, 2003].

The pair distribution function (PDF) obtained from the powder X-ray and neutron diffraction experiments have been shown to be of great value in determining the local atomic structure of the materials [Billinge and Thorpe, 1998]. The PDF results from a Fourier transform of the powder diffraction spectrum (Bragg peaks + diffuse scattering) into real space [Egami and Billinge, 2003]. For well-ordered crystals, apart from technical details, this is similar to fitting the Bragg peaks and thermal diffuse scattering in the powder pattern in a manner first discussed by Warren [1953]. A PDF spectrum consists of a series of peaks, the positions of which give the distances of atom pairs in real space. The ideal width of these peaks (aside from problems of experimental resolution) is due both to relative thermal atomic motion and to static disorder. Thus an investigation of the effects of lattice vibrations on PDF peak widths is important for at least two reasons. First to establish the degree to which information on phonons and the inter-atomic potentials can be obtained from powder diffraction data and second to account for correlation effects in order to extract information on static disorder in a disordered system such as an alloy.

In general, powder diffraction is not considered as a favorable approach for extracting information about phonons since, not only is energy information lost in the measurements, but also the diffuse scattering is isotropically averaged. The lattice vibration are best described from the phonon dispersion curves determined using inelastic neutron scattering and high energy resolution inelastic scattering on single crystals [Schwoerer *et al.*, 1998]. Nevertheless, with the advent of high energy synchrotron X-ray and pulsed neutron sources and fast computers, it is possible to measure data with unprecedented statistics and accuracy. The PDF approach has been shown to yield limited information about lattice vibrations in powders [Jeong *et al.*, 1999] though the extent of which this information can be extracted remains controversial [Dimitrov *et al.*, 1999; Reichardt and Pintschovius, 1999].

Measuring powders has the benefit that the experiments are straightforward and do not require single crystals. It is thus of great interest to characterize the degree to which lattice vibrations are reflected in the PDF using simple models, such as the Debye model, in situations where detailed inter atomic potential information is not available.

1.9.1 CORRELATED ATOMIC MOTION IN REAL SPACE

The existence of inter atomic forces in crystals results in the motion of atoms being correlated. This is usually treated theoretically by transforming the problem to normal coordinates, resulting in normal modes (phonons) that are non-interacting, thus making the problem mathematically solvable. Projecting the phonons back into real space coordinates yields a picture of the dynamic correlations. This situation can be understood intuitively in the following way.

In a rigid body system, inter atomic force is extremely strong and atoms move in phase. In this case, the peaks in the PDF are delta functions. At the opposite extreme the atoms are non-interacting (Einstein model) and move independently. This type of atomic motion is resulting broad PDF peaks, whose widths are given by the root mean square displacement amplitude ($\sqrt{< u^2 >}$). In real materials, inter atomic forces depend on atomic pair distances, i.e., they are strong for nearest-neighbor interactions and get weaker as the atomic pair distance increases. In fact, these interactions are often quite well described with just nearest-neighbor or first and second nearest-neighbor coupling.

In this model (Debye) a single parameter corresponding to the spring constant of the nearest-neighbor interaction is used. Here, near-neighbor atoms tend to move in phase with each other, while far neighbors move more independently. As a result, the near-neighbor peaks are sharper than those of far neighbor-neighbor pairs. This behavior was first analyzed by Kaplow and co-workers in a series of papers [Kaplow *et al.*, 1964, 1965; Lagneborg and Kaplow, 1967] for a number of elemental metals.

1.9.2 ALGORITHM OF PAIR DISTRIBUTION FUNCTION

The scattered X-ray intensity by a collection of atoms can be expressed as follows (After corrections for absorption, polarization, multiple scattering and normalization to the unit of one atom or scattering),

$$I(\vec{Q}) = \Sigma_{i,j} f_i\,(\vec{Q}) f_j(\vec{Q}) \ll \exp(i\vec{Q}.(r_i - r_j)) \gg. \qquad (1.44)$$

Where $f_j(Q)$ is the scattering amplitude of single atom i, r_i is the position of the ith atom, $\ll \exp\left(i\vec{Q}.(r_i - r_j)\right) \gg$ is the quantum and thermal average $\vec{Q}$ is given by,

$$\vec{Q} = K_f - K_i \qquad (1.45)$$

$$Q = |\vec{Q}| = 2|k_{f,i}|sin\theta \qquad if\ \vec{K}_f = \vec{K}_i \qquad (1.46)$$

Here $\vec{K_f}$ and $\vec{K_i}$ are the momenta of scattered and incident X-ray photons respectively and θ is the diffraction angle. The quasi-elastic X-ray scattering includes inelastic phonons scattering too.

The average structure factor can be given as,

$$S(\vec{Q}) = \frac{I(\vec{Q})}{<f(\vec{Q})>^2} + \frac{[<f(\vec{Q})>^2 - <f(\vec{Q})^2>]}{<f(\vec{Q})>^2} \quad (1.47)$$

Here $< f(\vec{Q}) >^2$ is the compositional average. This equation for $S(\vec{Q})$ can be noted as the square of the structure factor $F(\vec{Q})$.

Instead of indexing and analyzing each powder peak separately, the total data will be treated in real space, by Fourier transforming the data. The Fourier transform of equation (1.47) in 3D, gives the atomic Pair distribution function (PDF).

$$\rho(r) = \rho_0 + \frac{1}{2\pi^2 r}\vec{Q}[S(\vec{Q}) - 1]\sin(\vec{Q}r)\,d\vec{Q} \quad (1.48)$$

Where ρ_0 = Average (atomic) number density (average no of atoms in the sample with respect to distance) and r = distance.

If Q is in Å, then $\rho(r)$ corresponds to the (atomic) number density at a distance r from the average atom. Equation (1.48) is expressed as follows,

$$G(r) = 4\pi r[\rho(r) - \rho_0] = \frac{2}{\pi}\int \vec{Q}\,[S(\vec{Q}) - 1]\,\sin(\vec{Q}\,r)d\,\vec{Q} \quad (1.49)$$

The atomic pair distribution function can be obtained from powder diffraction data and is a valuable tool for the study of the local atomic arrangements in a material. The Bragg and diffuse scattering information about local arrangements is preserved in PDF. The PDF can be understood as a bond-length distribution between all pairs of atoms within the crystal (up to the maximum distance); however, each contribution has a weight corresponding to the scattering power of the two atoms involved.

In the present work, the powder X-ray intensity data of the thermoelectric materials Mg_2Si, PbTe, Sb_2Te_3 and Bi_2Te_3 have been collected and normalized to obtain the total scattering functions S(Q), using the software PDFgetX [Jeong *et al.*, 2001]. The experimental PDF, G(r), was obtained by taking the Fourier transform of S(Q) using equation (1.49). The experimental PDF peak widths as a function of pair distance are extracted using the "real space" Rietveld program PDFGUI [Farrow *et al.*, 2007].

1.10 A REVIEW OF THE EARLIER LITERATURE DONE ON THE MATERIALS STUDIED IN THIS BOOK

1.10.1 MAGNESIUM SILICIDE (MG_2Si)

The early works on Mg_2Si proved that it is an intermediate thermoelectric power generator material with maximum operating temperature of around 600 K and hence it plays an important role, in high temperature applications. The preparation and characterization of Mg_2Si has been reported in numerous research articles. For example, Lee *et al.,* [2006] has reported the preparation of Mg_2Si by mechanical alloying methods; Zang *et al.,* [2003] and Tani and Kido [2007] have studied the thermoelectric properties of Sb doped Mg_2Si; Song *et al.,* [2005] reported the synthesis and thermoelectric properties of $Mg_2Si_{1-x}Ge_x$ (x=0, 0.2, 0.4, 0.6, 0.8, 1); Bose *et al.,* [1993] have studied the thermoelectric properties of Mg_2Si; Kamilov *et al.,* [2006] reported the thin film preparation and characterization of Mg_2Si; Bashenov *et al.,* [1978] reported the calculation of density of states in Mg_2Si and Akasaka *et al.,* [2007] reported the growth of Mg_2Si using vertical Bridgman method and its characterization. A study on the local structure and charge distribution has not been reported in the literature on Mg_2Si.

1.10.2 LEAD TELLURIDE (PbTe)

Lead telluride (PbTe) is reported as a high temperature thermoelectric power generator material with maximum operating temperature of 900 K. PbTe has a high melting point, good chemical stability, low vapor pressure and good chemical strength in addition to high figure of merit *Z*. In this class of materials, PbTe alloys play an important role, in high temperature applications. The figure of merit is high in $AgPb_mSbTe_{m+2}$ alloys with ZT=2.2 at 800 K [Hsu *et al.,* 2004]. Commercially available thermoelectric materials have a variety of applications. In PbTe alloys, Pb can be replaced by Ag and/or Sb [Thompson *et al.,* 2006], but the exact role and distribution of alloying elements on the different sub lattices and the structure and composition of the clusters are still to be studied.

Recently, the study of thermoelectric materials has once again become an active research field, in part due to the recent demonstration of enhancement in the thermoelectric figure of merit of a two-dimensional PbTe quantum well system, relative to its three-dimensional (3D) bulk counterpart [Hicks *et al.,* 1996]. Calculations suggest that the thermoelectric performance of any 3D material should show an enhanced thermoelectric figure of merit, when prepared as a 2D multi quantum well super lattice, utilizing the enhanced density of states at the onset of each electronic sub band, and the increased scattering of vibrational waves at the boundary between the quantum well and the adjacent barrier of the super lattice.

Many studies are available on the disordered thermoelectric structures, an example being the high temperature high pressure processing of PbTe with KCl to produce porous PbTe [Hideyuki *et al.,* 1999]. Although the efficiency of thermoelectric generators is rather low, which is nearly 5%, the other advantages, such as compactness, silent, reliability, long life, and long period of operation without attention, led to a wide range of applications. PbTe thermoelectric generators have been widely used by the US army, in spacecrafts to provide onboard power, and in pacemaker batteries. The general physical properties of lead telluride and factors affecting the figure of merit have been reviewed by many researchers. Various possibilities of improving the figure of merit of the material have been given, including effect of grain size on reducing the lattice thermal conductivity K_{lat}. The literature survey on this material reveals no information regarding the electron level properties like the electron density inside the material and also the bond length distribution. Hence, this material is chosen for the charge density analysis in this work.

1.10.3 BISMUTH DOPED WITH ANTIMONY ($Bi_{80}Sb_{20}$)

Several binary and ternary compounds containing bismuth (Bi) antimony (Sb) and Tellurium (Te) were reported to be good thermoelectric materials with high figure of merit (ZT). For example, it is reported that the binary compound Bi_2Te_3 has a maximum figure of merit of 0.6 at 300K [Sharp *et al.,* 1999] and the ternary material $CsBi_4Te_6$ with ZT of 0.8 at 225K [Chung *et al.,* 2000]. $Bi_{1-x}Sb_x$ is reported to have low effective mass and high mobility of charges with antimony atoms iso-electronic with bismuth with the same structure. The electronic properties can be varied by the variation of Sb concentration. It is further reported that it has a semi metallic behavior, when x is below 0.07. It is an indirect band gap semiconductor in the interval of x between 0.07 to 0.22 and direct band gap semiconductor in the interval ($0.09 < x < 0.16$).

In bulk materials, new synthesis routes lead to complex crystal structures with the desired phonon-glass electron- crystal (PGEC) behavior [Slack, 1995]. These results have led to intensified scientific efforts to achieve high figure-of-merit, much larger than one near room temperature. This PGEC behavior needs to have a complete structural analysis of the material in terms of electron density distribution of both core and valance electrons, thermal vibration parameter, atomic size and bond length distribution. Hence, the present work is aimed at the structural analysis in terms of the local and average structural properties like the electron density distribution and bonding between atoms of the thermoelectric materials $Bi_{80}Sb_{20}$, since this material is reported as good thermoelectric material with reasonable figure of merit [Cho *et al.,* 1999]. Though this material is reported as a good thermoelectric material, the microscopic analysis like the charge density analysis and the inter-atomic correlation study will reveal much more information, supporting its thermoelectric behavior.

1.10.4 BISMUTH TELLURIDE AND ANTIMONY TELLURIDE (Bi_2Te_3 AND Sb_2Te_3)

The early work on thermoelectric research in the 1950's and early 1960's showed that Bi_2Te_3 is a high potential thermoelectric material which was discovered by H.J. Goldsmid and coworkers in U.K. [Goldsmid *et al.,* 1954]. These systems and their solid solutions remain the basis for the thermoelectric industry up to the present time [Goldsmid, 1986]. The semiconductor alloys based on Bi_2Te_3, Sb_2Te_3 and Bi_2Se_3 family have proved their usefulness in industrial applications [Ioffe *et al.,* 1956]. The best commercial bulk thermoelectric material is reported to be $Bi_{0.5}Sb_{1.5}Te_3$ [Goldsmid, 1986] with ZT=1. It is also reported that Bi_2Te_3/Sb_2Te_3 is a narrow band gap semimetal alloy with homologous layered crystal structure having five atoms per unit cell [Kim *et al.,* 2002]. It is also reported that the superlattice Bi_2Te_3/Sb_2Te_3 exhibit a high thermoelectric figure of merit of ZT= 2.4. [Venkatasubramaniun *et al.,* 2001]. A 30% higher figure of merit, (ZT= 1.17) has been reported for PbTe alloyed with 0.8 mol% Sb_2Te_3, at 560 K than conventional bulk PbTe [Pinwen *et al.,* 2005]. It is recently reported that the Bi_2Te_3-Sb_2Te_3 bilayer thin film has a figure of merit 0.49 compared to Sb_2Te_3 and Bi_2Te_3 films which are 0.17 and 0.43 respectively. [Pradyumnan and Swathikrishnan, 2009]. The thermal conductivity studies in Sb_2Te_3-Gd_2Te_3-Bi_2Te_3 have shown that the heat transported through lattice phonons are the key to the thermoelectric property. (Ismaiyova *et al.,*2009). The thermoelectric figure of merit and power factor of Ti_2Te-Sb_2Te_3 pseudo binary system were reported as 0.42 at 591 K and 6.53×10^{-4} $Wm^{-1}K^{-2}$ for Ti_9SbTe_6 [Ken *et al.,* 2006]. An effort has been made in this work to study the structure and electron density distribution of Bi_2Te_3 and Sb_2Te_3.

1.10.5 TIN TELLURIDE DOPED WITH GERMANIUM ($Sn_{1-x}Ge_xTe$)

The experimental parameters like Seebeck coefficient, thermoelectric figure of merit and thermoelectric power of SnTe and their solid solutions have been reported by many researchers. The thermoelectric power of p-type SnTe has been reported between room temperature and 450°C [Skrabek and Trimmer, 1995]. Also the TAGS-85, i.e., $(AgSbTe2)_{0.15}(GeTe)_{0.85}$-SnTe segmented p-type thermoelectric generator has been used successfully on several NASA space missions [Jeffrey synder and Caillat, 2004]. A phase transition study of germanium telluride (GeTe) was reported by Korzhuev [1982]. It was reported that this material undergo a phase transition from cubic to rhombohedral structure near the temperature range 625 K and 690 K. $Pb_{1-x}Sn_xTe$, $Pb_{1-x}Ge_xTe$ and $Pb_{1-x}Ge_xS$ are some important and well-studied IV–VI systems as far as the phonon properties, thermal properties and phase transition studies are concerned (Fano *et al.*, 1977). The optical and thermal properties of the mixed semiconducting alloy, $Sn_{1-x}Ge_xTe$, were reported for various Ge concentrations [Sivabharathi *et al.*, 2005]. The behaviour of electrical resistivity and X-ray microanalysis of SnTe-SnSe quaternary semiconductor solid solutions at low temperatures, 0.4-4.2 K were reported [Moshnikov *et al.*, 1999]. An attempt was made to analyse the electron density distribution in $Sn_{1-x}Ge_xTe$ (x =0.12 and 0.25) in the present work, because the electron density distribution and the bonding nature of these thermoelectric materials are not available in current literature. The present work on $Sn_{1-x}Ge_xTe$ can be considered to be a clear and precise attempt in visualizing the electron density distribution and bonding nature in the thermoelectric material $Sn_{1-x}Ge_xTe$ using single crystal X-ray data.

1.10.6 INDIUM ANTIMONIDE (InSb)

The emerging families of advanced thermoelectric materials are dominated by antimonides and tellurides. Because the structures of the tellurides are mostly composed of NaCl-related motifs, they do not contain any Te-Te bonds, and all of the antimonide structures exhibit Sb-Sb bonds of various lengths [Jianxiao and Kleninke, 2008]. It was reported that the InSb nanowire grown using a vapor-liquid-solid method has better control of the impurity doping concentration, which improves the thermoelectric properties [Jae *et al.*, 2007]. InSb is mainly used as infrared detectors and Hall sensors [Al-Ani *et al.*, 2009]. But recent studies have shown that InSb has much potential thermoelectric device applications [Matsumoto *et al.*, 2007]. It is reported that the Seebeck coefficient and power factor for Si doped and undoped InSb thin films are -214μV/K and -261μV/K and $2x10^{-4}W/mK^2$ and $1.3x10^{-3}W/mK^2$ respectively. Further, the thermoelectric properties of GaSb and InSb alloys were studied and reported [Qiang Wang *et al.*, 2000]. The material InSb, both in bulk and thin film form is potentially useful for device applications. Therefore, an attempt has been made in this work to study the distribution of charge density using the single crystal X-ray data.

1.11 PRESENT BOOK

From the review of literature, it was found that commercially available thermoelectric materials have a variety of applications, but the efforts to understand the underlying mechanisms by which the local structure affects the performance of thermoelectric materials are still going on. Though there is much experimental work on the growth and physical characterization of thermoelectric materials, only limited information about the local structure, electron density distribution and bonding is available.

Hence, the present work was aimed at the exploration of the fine structural details of some of the conventional thermoelectric materials, with careful attention in both experiment and computations. The nature of the work motivated the author to take maximum care for the deep analysis of the crystal structures like local bond length distribution analysis and electron density analysis rather than the surface level studies. The author has made every effort to extract the maximum information possible from the simple and precise X-ray diffraction data both in single crystal form as well as powder crystal form.

The materials chosen for the present analysis are the conventional bulk thermoelectric materials Mg_2Si, PbTe, $Bi_{1-x}Sb_x$, Bi_2Te_3, Sb_2Te_3, $Sn_{1-x}Ge_xTe$ and InSb. The materials Mg_2Si, PbTe, $Bi_{1-x}Sb_x$, Bi_2Te_3 and Sb_2Te_3 were studied using powder X-ray diffraction data and the materials $Sn_{1-x}Ge_xTe$ and InSb using single crystal X-ray method. The powder X-ray diffraction data of the samples Mg_2Si, PbTe, Bi_2Te_3 and Sb_2Te_3 were recorded from the analar grade (99.999%) chemicals purchased from Alfa Acer, Johnson's Mathew company. The remaining materials were grown using suitable methods and in suitable combinations. Further, the materials $Sn_{1-x}Ge_xTe$ and InSb were analysed in single crystal form.

The entire work done can be classified as (i) Electron density analysis and (ii) local structural analysis (bond length distribution analysis). Altogether, both roots explore the maximum information for the fruitful analysis of the thermoelectric behavior of the selected materials.

1.11.1 PRESENT WORK ON ELECTRON DENSITY ANALYSIS

The single crystal as well as powder crystal data recorded from suitable X-ray diffractometer was indexed with possible miller indices planes. The structure factor for various possible miller indices (h k l) were deduced and compared with observed ones using the standard Rietveld technique with the help of the software JANA2006 [Petříček *et al.*, 2006]. This process also yields various crystallographic parameters like the cell parameters, thermal vibration parameters and atomic position of the atoms etc.

The refined structure factors were used along with other details like space group atomic positions etc., using the software PRIMA [Izumi and Dilanian, 2002] to elucidate the fine details of the electron distribution inside the unit cell using the versatile technique maximum entropy method (MEM). The 3D, 2D electron densities inside the unit cell on various planes and directions were perfectly visualized and the numerical one dimensional electron densities were calculated and plotted using the software program VESTA (Visualization for Electronic and Structural Analysis) [Momma and Izumi, 2006]. This provides a clear picture of the strength and nature of bonding between the atoms inside the unit cell of the selected thermoelectric materials.

1.11.2 PRESENT WORK ON LOCAL STRUCTURAL ANALYSIS (BOND LENGTH DISTRIBUTION ANALYSIS)

The second root of analysis of thermoelectric materials was the real space bond length distribution analysis using pair distribution function (PDF) [Billinge, 1999]. This technique uses the background part of the X-ray powder diffraction data (diffuse scattering), in addition to the Bragg reflections. Much of the information can be retrieved using this technique, which cannot be retrieved using other techniques.

PDF is the instantaneous atomic number density - density correlation function which describes the atomic arrangement in materials. A useful characteristic of PDF method is that it gives both local and average structure information because both Bragg peaks and diffuse scattering are used in the analysis. It is also possible to obtain information about bond length distribution (static, thermal) [Petkov *et al.*, 1999] and correlated atomic thermal motion [Jeong *et al.*, 2002] using the PDF peak width.

The atomic pair distribution function (PDF) is obtained from the Fourier transform of the measured X-ray powder diffraction data using the software PDFgetX [Jeong *et al.*, 2001]. Many corrections have been carried out for experimental effects such as absorption, polarization and removing of Compton and multiple scattering contributions to the elastic scattering during the process of obtaining total structure function and PDF.

The observed and calculated PDF were compared using the graphical software PDFgui [Farrow *et al.*, 2007], which is a graphical environment for PDF fitting. This program has powerful usability features such as real time plotting and remote execution of the fitting program whilst visualizing the results locally.

During the PDFfit [Proffen and Billinge, 1999] refinement, the structural parameters like, lattice parameters, phase scale factor, linear atomic correlation factor, quadratic atomic correlation factor, spherical nano particle amplitude correction, low r peak sharpening, peak sharpening cut off and cutoff for profile setup functioning were refined to get the absolute phase. The data configuration parameters are PDFfit range with step size, data scale factor, upper limit for Fourier transform to obtain data PDF, resolution damping factor, resolution peak broadening factor, data collection temperature and doping concentration levels etc., which can be refined to get accurate PDF fitting.

Finally, all the numerical and pictorial information was consolidated with theoretical prediction for the better understanding of the materials chosen.

REFERENCES

[1] Abelson R.D, 'Space Missions and Applications' Thermoelectric: Macro to Nano, D.M. Rowe Ed. CRC Press (2005).

[2] Akasaka M, Iida T, Nemoto T, Soga J, Kato J, Makino K, Fukano M and Takanashi Y, J. Cryst. Growth, Vol. 304 (2007) p. 196. *http://dx.doi.org/10.1016/j.jcrysgro.2006.10.270*

[3] Al-Ani S.K.J, Obaid Y.N, Kasim S.J and Mahdi M.A, Int.J. Nano electronics and Materials Vol. 21 (2009) pp. 99-109.

[4] Allen P.B and Feldman J.L, Phys. Rev. Vol. B48 (1993) p. 12581. *http://dx.doi.org/10.1103/PhysRevB.48.12581*

[5] Androulakis and Pantelis Migiakis, Applied physics letters Vol. 84 (2004) 1099. *http://dx.doi.org/10.1063/1.1647686*

[6] Ashcroft N. W and Mermin N. D, Solid State Physics, Philadelphia, Saunders (1976).

[7] Bashenov V. K, Mutal A.M and Timofeenko V.V, Phy. Stat. Solidi (b) Vol. 87 (1978) p. K77. *http://dx.doi.org/10.1002/pssb.2220870247*

[8} Bass J, Elsner N.B and Leavitt F.A, Proc. 13th Int.Conf.Thermoelectrics, Ed. Mathiprakisam B., AIP Conf. Proc. New York, (1995) p. 295.

[9] Bennett G,'Space Applications' Chapter 41, CRC Handbook of Thermoelectric, Ed. D.M. Rowe, (2005) CRC Press.

[10] Billinge S and Thorpe M.F, Local structure from Diffraction, Plenum, New York, (1998).

[11] Bose S, Acharya H.N, Banerjee H.D, J. Mat. Science Vol. 28 (1993) p.5461. *http://dx.doi.org/10.1007/BF00367816*

[12] Bourenkov G.P, Popov A.N and Bartunik H.D, Acta Cryst. Vol. A52 (1996) pp. 797-811. *http://dx.doi.org/10.1107/S0108767396005648*

[13] Bowran D.T, Finney J.L, J. Chem. Phys., Vol. 118 (2003) p. 8357. *http://dx.doi.org/10.1063/1.1565102*

[14] Bragg W.L. "The Diffraction of Short Electromagnetic Waves by a Crystal", Proceedings of the Cambridge Philosophical Society Vol. 17 (1913) p. 43.

[15] Brebrick R.F and Strauss A.J, Phys.Rev. Vol. 131 (1963) pp.104-110. *http://dx.doi.org/10.1103/PhysRev.131.104*

[16] Brian, R.M. & Cohen, R.S. (2007). Hans Christian Ørsted and the Romantic Legacy in Science, Boston Studies in the Philosophy of Science Vol. 241. *http://dx.doi.org/10.1007/978-1-4020-2987-5*

[17] Cagliot G, Paolett A and Ricci F.P, Nuclear Instrum. Vol. 3 (1958) pp.223-228. *http://dx.doi.org/10.1016/0369-643X(58)90029-X*

[18] Caillat T, Borshchevsky A, and Fleurial J.P, J. Appl. Phys. Vol. 80 (1996) pp. 4442-4449. *http://dx.doi.org/10.1063/1.363405*

[19] Callen H.B, Thermodynamics and an introduction to thermo statistics 2nd ed., Chapter 14, New York, John Wiley & Sons, (1985).

[20] Chen G, Phys. Rev. B, Vol. 57-23 (1998) p.14958.

[21] Chen L.D, Kawahara T, Tang X.F, Goto T and Hirai T, Journal of applied physics Vol. 90, (2001) p. 1864. *http://dx.doi.org/10.1063/1.1388162*

[22] Cho S, DiVenere A, Wong G.K, Ketterson J.B, Meyer J.R and Hoffman C.A, Phys. Rev. B Vol. 59, (1999) pp. 10691–10696. *http://dx.doi.org/10.1103/PhysRevB.59.10691*

[23] Chung D.Y, Hogan T, Brazis P, Rocci-Lane M, Kannewurf C, Bastea M, Uher C, and Kanatzidis M.G, Thermoelectrics hand book: macro to nano, Vol. 287 (2000) p. 1024-1027.

[24] Cohn J.L, Nolas, G.S, Fessatidis, V. Metcalf, T.H and Slack G.A, Phys. Rev. Lett. Vol. 82 (1999) p. 779. *http://dx.doi.org/10.1103/PhysRevLett.82.779*

[25] Collins D.M, Nature Vol. 298 (1982) p. 49. *http://dx.doi.org/10.1038/298049a0*

[26] Cox D.E, Hastings J.B, Thomlinson W and Prewitt C.T, Nucl. Instrum. Methods. Vol. 208, (1983) pp. 573-578. *http://dx.doi.org/10.1016/0167-5087(83)91185-7*

[27] Debay P and Menki H, Physika Zeit Vol. 31 (1930) p. 797.

[28] Dimitrov D.A, Louca D, and Roder H, Phys. Rev. Vol. B60 (1999) p. 6204.

http://dx.doi.org/10.1103/PhysRevB.60.6204

[29] Egami T and Billinge S.J.L., "Underneath the Bragg Peaks: Structural Analysis of Complex Material", Oxford University Press, London (2003).

[30] Ellis A.B, Geselbracht M.J Johnson B.J, Lisensky G.C and Robinson W.R.A Materials Science Companion; ACS Books: Washington, DC, (1993).

[31] Fan X.A, Yang J.Y, Chen R.G, Yun H.S, Zhu W, Bao S.Q and Duan X.K, J. Phys. D: Appl. Phys. Vol. 39 (2006) p. 740.
http://dx.doi.org/10.1088/0022-3727/39/4/021

[32] Fano V, Fedeli G and Ortali I, Solid State Commun. Vol. 22 (1977) p. 467.
http://dx.doi.org/10.1016/0038-1098(77)90127-2

[33] Farrow C.L, Juhás P, Liu J.W, Bryndin D, Bozin E.S, Bloch J, Proffen Th and Billinge S.J.L, "PDFfit2 and PDFgui: Computer programs for studying nanostructure in crystals", J. Phys.: Condens. Matter Vol. 19 (2007) p. 335219.
http://dx.doi.org/10.1088/0953-8984/19/33/335219

[34] Federov M and Zsitsev Z, Thermoelectric Handbook, Macro to Nano, D.M.Rowe, CRC press, Boca Raton, USA (2005).

[35] Fujita K, Mochida T, and Nakamura K, J. Appl. Phys. Vol. 40 (2001) p. 4644-1617.
http://dx.doi.org/10.1143/JJAP.40.4644

[36] Funahashi R, Matsubara I, Ikuta H, Takeuchi T, Mizutani U, and Sodeoka S, Jpn. J. Appl. Phys. Vol. 39 (2000) pp. L1127-1129.
http://dx.doi.org/10.1143/JJAP.39.L1127

[37] Funahashi R and Matsubara I, Appl. Phys. Lett, Vol. 79 (2001) pp. 362-364.
http://dx.doi.org/10.1063/1.1385187

[38] Geetham A.K and Taylor J.C, J. Solid State Chem. Vol. 21(1977) pp. 253-75.

[39] Gilmore C.J., Acta Cryst. Vol. A52, (1996) p. 561.
http://dx.doi.org/10.1107/S0108767396001560

[40] Goldsmid H.J and Douglas R.W, Brit. J. Appl. Phys. Vol. 5 (1954) pp. 386, 458.

[41] Goldsmid H. J. Electronic Refrigeration, Pion, London (1986) p. 10.

[42] Hebert S, Lambert S, Pelloquin D, and Maignan A, Phys. Rev. Vol. B64 (2001) p. 172101.
http://dx.doi.org/10.1103/PhysRevB.64.172101

[43] Heikes R.R and Ure R.W, Jr. Thermoelectricity: Science and Engineering, Interscience, New York (1961) p. 20.

[44] Hicks L.D, Harman T.C, Sun X and Dresselhaus M.S. Phys. Rev. Vol. B53 (1996) p. 10493.
http://dx.doi.org/10.1103/PhysRevB.53.R10493

[45] Hideyuki Y, et al., Journal of the Japan Institute of Metal Vol. 63, 11 (1999) p. 1468.

[46] Hsu K, Loo S and Guo F, Science Vol. 303 (2004) p. 818.
http://dx.doi.org/10.1126/science.1092963

[47] International Tables for X-ray Crystallography, The Kynoch press, Birmingam, England, (1974).

[48] International Tables for Crystallography, Vol. A. Th. Hahn, editor. Reidel, Dordrecht, Holland (1993).

[49] Ioffe A.F, Semiconductor Thermo-elements and Thermoelectric cooling, info search, London (1957).

[50] Ichiro Terasaki, Introduction to thermoelectricity, Department of Applied Physics, Waseda University, Ichiro Terasaki, chapter-10 (2005) p. 3.

[51] Ioffe A.F, Airapetyants S.V, Ioffe A.V, Kolomoets N.V and Stilbans L.S, Dokl. Akad. Nauk SSSR Vol. 102, (1956) p. 981.

[52] Ismaiyova R.A, Bakhtiyarly I.B and Abdinov D. Sh, Neorganicheskie Materialy Vol. 45, 7 (2009) pp. 806-809.

[53] Izumi F and Dilanian R.A, Recent Research Developments in Physics Vol. 3, Part II, Transworld Research Network, Trivandrum (2002) pp. 699–726.

[54] Jae Hun Seol, Arden L. Moore, Sanjoy K. Saha, Feng Zhou, and Li Shi, J. Appl. Phys. Vol. 101 (2007) p. 023706.
http://dx.doi.org/10.1063/1.2430508

[55] Jeffrey Snyder G, Jet Propulsion Laboratory, California Institute of Technology (2004).

[56] Jeffrery Synder G, and Caillat T, MRS proceedings Vol. 793 (2004) p. 37.

[57] Jeong L.K, Proffen Th, Mohiuddin Jacobs F, and Billinge S.J.L, J. Phys. Chem. Vol. A 103 (1999) p. 921.
http://dx.doi.org/10.1021/jp9836978

[58] Jeong K, Thompson J, Proffen TH, Perez A and Billinge S.J.L, "PDFGetX, A Program for Obtaining the Atomic Pair distribution function from X-ray Powder Diffraction Data", (2001).

[59] Jianxiao Xu and Holger kleninke, Department of chemistry, University of Waterloo, Waterloo, Ontario N2L 3G1, Canada (2008).

[60] Kamilov T.S, Kabilov D.K, Kamilova R.Kh, Azimov M.E, Kleechkovskaya V.V, Orekhov A.S and Suvorova E.I, Thermoelectrics 25nd International Conf. on Vol. ICT (2006) p. 468.

[61] Kaplow R, Averbach b.L, andStrong S.L, J. Phys. Chem. Solids Vol. 25 (1964) p. 1195.
http://dx.doi.org/10.1016/0022-3697(64)90016-2

[62] Kaplow R, Strong S.L. and Averbach B.L, Phys. Rev. Vol. 138 (1965) p. 1336.
http://dx.doi.org/10.1103/PhysRev.138.A1336

[63] Ken Kurosaki, Keita Goto, Atsuko Kosuga, Hiroaki Muta and Shinsuke Yamanak Thermoelectric Society of Japan, Conference on 3rd Proceedings (2006) pp. 36-37.

[64] Kim Y, DiVenere A, Wong G.K.L, Kelterson J.B, Cho S and Meyer J.R, J. Appl. Phys. Vol.91 (2002) p. 715.
http://dx.doi.org/10.1063/1.1424056

[65] Koeningsberger D.C and Prins R, X-ray Absorption, principles, applications and techniques of EXAFS, SEXAFS and XANES, edited by Wiley (1988).

[66] Korzhuev M.A, Phys. Status Solidi Vol. (b) 112 (1982) p. K39.
http://dx.doi.org/10.1002/pssb.2221120149

[67] Koumoto K, Terasaki I and Murayama N, Oxide Thermoelectrics, Trivandrum, Research Signpost, Ed. (2003).

[68] Kumazawa S, Kubota, Takata M and Sakata, J. Apll. Cryst. Vol. 26 (1995) p. 453-457.
http://dx.doi.org/10.1107/S0021889892012883

[69] Kumazawa S, Takata M and Sakata M, J. Acta. Cryst. Vol. A51 (1995) pp. 47-53.
http://dx.doi.org/10.1107/S0108767394006173

[70] Kyoung Jeong-II, Thomas Proffen, Farida Mohiuddin-Jacobs and Simon J. L. Billinge, J. Phys. Chem. Vol. A 1999, p. 103, 921-924.

[71] Lagneborg and Kaplow, Acta Metall. Vol .15 (1967) p. 13.
http://dx.doi.org/10.1016/0001-6160(67)90150-2

[72] Lee C. H, Lee S.H, Chun S.Y and Lee S.J, J. Nano science and Nanotechnology Vol .6 (2006) p. 3429.

[73] Mahan G.D, Figure of merit for thermoelectric Appl. Phs. Vol. 65 (1989) pp. 1578-1583.

[74] Mahan G.D, Good thermoelectrics Solid State Phys. Vol. 51 (1998) p. 81-157. *http://dx.doi.org/10.1016/S0081-1947(08)60190-3*

[75] Maignan A, Hebert S, Pelloquin D, Michel C and Hejtmanek J, Appl. Phys. Vol. 92 (2002) pp. 1964-1967. *http://dx.doi.org/10.1063/1.1494114*

[76] Malmros G and Thomas J.O, J. Appl. Crystallogr. Vol. 10 (1977) p. 11. *http://dx.doi.org/10.1107/S0021889877012680*

[77] Matsubara, 21st international conference on Thermoelectrics, proceedings ICT Vol. 2, (2002) p. 418.

[78] Matsumoto M, Yamazki J, Yamaguchi S, Mater. Res. Soc Symp. Proc. Vol. 0980-II05-42 (2007).

[79] Momma K and Izumi F, Commission on Crystallogr. Comput. IUCr Newsle Vol 7 (2006) p. 106.

[80] Moshnikov and Rumyantseva, Physics of The Solid State, Vol. 41 (1999) pp.612-617.

[81] Nolas G.S, Cohn J.L, Slack G.A and Schujman S.B, Appl. Phys. Lett. Vol .73 (1998) pp. 178-180. *http://dx.doi.org/10.1063/1.121747*

[82] Onsager L, Reciprocal Relations in Irreversible Processes I, Physical Review Vol. 37 (1931) pp. 405-426. *http://dx.doi.org/10.1103/PhysRev.37.405*

[83] Papoular R.J and Cox D.E, Europhys. Lett. Vol. 32 (1995) p. 337. *http://dx.doi.org/10.1209/0295-5075/32/4/009*

[84] Papoular R.J, Cousson A, Paulus W and Kaiser-Morris E, Physica B Condensed Matter Vol. 234-236 (1997) pp. 72-73. *http://dx.doi.org/10.1016/S0921-4526(96)00884-8*

[85] Paul E. Gray, The Dynamic behavior of thermoelectric devices, Published jointly by the Technology press of the Massachusetts Institute of technology and John Wiley and sons, Inc., New York, (1960).

[86] Peltier, J.C. (1834). Nouvelles experiences sur la caloricité des courans électriques. Annales de Chimie et the Physique, Vol. LVI 56, pp. 371-386.

[87] Petkov V, Trikalitis P.N, Bozin E.S, Billinge S.J.L, Vogt T and Kanatzidis M.G, J. Am. Chem. Soc. Vol. 124 (2002) pp. 10157–10162. *http://dx.doi.org/10.1021/ja026143y*

[88] Petricek P, Dusek M, and Palatinus L, JANA 2006, The crystallographic computing system, Institute of Physics, Academy of sciences of the Czech republic, Praha, (2006).

[89] Pinwen Zhu, Yoshio Imai, Yukihiro Isoda and Yoshikazi Shinohara, J. Phys: Condens. Matter Vol. 17 (2005) p. 7319. *http://dx.doi.org/10.1088/0953-8984/17/46/015*

[90] Pradyumnan P.P and Swathikrishnan, Indian Journal of Pure & Applied Physics Vol. 48 (2009) p. 115-120.

[91] Proffen T, Billinge S J L, PDFFIT, a Program for Full Profile Structural Refinement of the Atomic Pair distribution function, J. Appl. Cryst. Vol. 32 (1999) p. 572. *http://dx.doi.org/10.1107/S0021889899003532*

[92] Qiang Wang, Xiumei Chen and Kunquan Lu. Phys.: Condens. Matter Vol. 12 (2000) pp. 5201-5207.

[93] Reichardt W and Pintschovius L, Phys. Rev. Vol. B 63 (1999) p. 174302. *http://dx.doi.org/10.1103/PhysRevB.63.174302*

[94] Rietveld H.M, J. Appl. Cryst. Vol. 2 (1969) p. 65. *http://dx.doi.org/10.1107/S0021889869006558*

[95] Riffat S.B, Ma X. Thermoelectrics: A review of present and potential applications. Appl. Therm. Eng. Vol. 23 (2003) pp. 913-935. *http://dx.doi.org/10.1016/S1359-4311(03)00012-7*

[96] Rowe D.M, Thermoelectric waste heat recovery as a renewable energy source. Int. J. Innov. Energy Syst. Power, Vol. 1 (2006) pp. 13-23.

[97] Rowe D.M, Kuznetsov V.L, Kuznetsova L.A and Min G, J.Phys. D: Appl. Phys. Vol. 35 (2002) pp. 2183-2186. *http://dx.doi.org/10.1088/0022-3727/35/17/315*

[98] Rowe D.M and Min G, Proc. of 13th IT conference Kansas, USA (1994) p. 339.

[99] Saiki S, Takeda S.I, Onuma Y and Kobayashi M, Electrical Engineering in Japan, Vol. 105 (1985) p. 387.

[100] Sakata M and Toraya H, Acta. Cryst. Vol. A46 (1990) pp. 263-270.
http://dx.doi.org/10.1107/S0108767389012377

[101] Sakata.M, Uno T, Takata M and Mori R, Acta Crystallogr. Sect. A: Found. Crystallogr. Vol. A39 (1992) p. 47.

[102] Sales B.C, Mandrus D, Chakoumakos B.C, Keppens V, and Thompson V.R, Electron crystals and phonon glasses, Phys. Rev. Vol. B56 (1997) pp. 15081-15089.
http://dx.doi.org/10.1103/PhysRevB.56.15081

[103] Sales B.C, Mandrus D and Williams R.K, Science Vol. 272 (1996) p.1325.
http://dx.doi.org/10.1126/science.272.5266.1325

[104] Sales B.C, Chakoumakos B.C and Mandrus D, Phys. Rev. Vol. B 61 (2000) p. 2475.
http://dx.doi.org/10.1103/PhysRevB.61.2475

[105] Schleger P, Puig-Molina A, Ressouche E, Rutty O and Schweizer, J, Acta. Cryst. Vol. A53(1997) p. 426.
http://dx.doi.org/10.1107/S0108767397002158

[106] Schwoerer Bohning M, Macrander A.T and Arms D, Phys., Rev.Lett. Vol. 80 (1998) p. 5572.
http://dx.doi.org/10.1103/PhysRevLett.80.5572

[107] Seebeck, T. I. Magnetische polarisation der metalle und erze durch temperature differenz. Abhandlungen der Deutschen Akademie der Wissenschaften zu Berlin (1822) pp. 265-373.

[108] Sharp J.W, Sales B.C, Mandrus D.G and Chakoumakes B.C, App. Phys. Lett. Vol. 74/25 (1999) p. 3794.
http://dx.doi.org/10.1063/1.124182

[109] Sivabharathy M, Natarajan S, Ramakrishnan S.K and Ramachandran K, Bull. Mater. Sci. Vol. 27(2004) pp. 403.
http://dx.doi.org/10.1007/BF02708555

[110] Sivia D.S and David W, Acta Crystallographica Vol. A50 (1994) p. 70.

[111] Skilling J, Maximum Entropy and Bayesian methods. Kluwer Academic, London (1989).
http://dx.doi.org/10.1007/978-94-015-7860-8

[112] Skrabek E.A and Trimmer D.S, Thermoelectric Handbook, edited by Rowe D.M, CRC, Boca Raton, FL (1995) p. 267.

[113] Slack G.A, CRC Handbook of Thermoelectrics' edited by Rowe D.M, Chap. 34, Boca Raton FL,CRC Press, (1995).

[114] Slack G.A. "New Materials and Performance Limits for Thermoelectric Cooling" in CRC Handbook of Thermoelectrics, edited by Rowe D.M. (1995) p. 407.

[115] Song R, Yazheng L and Tatsuhiko A, J. Mat. Science and Technol. Vol. 21 (2005) p. 618.
http://dx.doi.org/10.1179/174328405X27025

[116] Spitzer D. P, Lattice Thermal Conductivity of Semiconductors: A Chemical Bond Approach, Phys. Chem. Solids Vol. 31 (1970) p. 19.
http://dx.doi.org/10.1016/0022-3697(70)90284-2

[117] Stout G.H, and Jensen L.H, X-ray structure determination, The Macmillan company, Collier – Macmillan Ltd., IV edition, London, (1970).

[118] Takata M, Sakata M, Kumazawa S, Larsen F, and Iversen B, Acta. Cryst. Vol. A50 (1994) p. 330
http://dx.doi.org/10.1107/S0108767393011523

[119] Tani J.I and Kido H, Intermetallics, Vol. 15 (2007) p. 1202.
http://dx.doi.org/10.1016/j.intermet.2007.02.009

[120] Terasaki I, Sasago Y and Uchinokura K, Phys. Rev. Vol. B56 (1997) pp. R12685-12687.
http://dx.doi.org/10.1103/PhysRevB.56.R12685

[121] Thompson K, Lawrence D and Larson D.J, Ultramicroscopy Vol. 107 (2-3) (2006) p. 131.
http://dx.doi.org/10.1016/j.ultramic.2006.06.008

[122] Venkatasubramaniun R, Siivola E, Colpitts T, and O'Quinn B, Nature Vol. 413597 (2001).

[123] Von Dreele R.B. and Geetham A.K. (1974), Proc. Roy. Soc. Vol. A338 (1974) pp. 311-326.
http://dx.doi.org/10.1098/rspa.1974.0088

[124] Von Dreele R.D., Jorgensen, J.D. and Windsor C.G, J. Appl. Crystallogr. Vol. 15, (1982) pp. 581-589.
http://dx.doi.org/10.1107/S0021889882012722

[125] Warren B.E, Acta crystallogr. Vol. 6 (1953) p. 803.
http://dx.doi.org/10.1107/S0365110X53002271

[126] Yamashita O and Sugihara S, Journal of Material Science, Vol. 40 (2005) pp. 6439-6444.
http://dx.doi.org/10.1007/s10853-005-1712-6

[127] Yodovard P, Khedari J, Hirunlabh J. The Potential of Waste Heat Thermoelectric Power Generation From Diesel Cycle and Gas Turbine Cogeneration Plants, Energy Sources Vol. 23 (2001) pp. 213-224.
http://dx.doi.org/10.1080/00908310151133889

[128] Young R.A, Mackie P.E and Von Dreele R.B, J. Appl. Crystallogr. Vol. 10 (1977) p. 2629.

[129] Zang L.M, Wang C.B, Jiang H.Y, Shan Q, Thermoelectrics 22nd International Conf. on ICT Vol. 146 (2003).

CHAPTER II

Results and Discussion Based on Rietveld Refinements

Abstract

The complete understanding of thermoelectric behavior needs the knowledge of the complete structural details of thermoelectric materials. The study of the structural details, particularly, the charge distribution of thermoelectric materials is not yet reported to a satisfactory level. The structural properties are of great importance in understanding the figure of merit in thermoelectric materials. This chapter provides the average structural analysis of selected thermoelectric materials using the software JANA2006 [Petříček *et al.*, 2006] and employing the Rietveld technique [Rietveld, 1969]. This chapter also reveals the results and discussion based on the Rietveld [Rietveld, 1969] analysis, particularly in terms of cell parameters, thermal vibration parameters etc. More detailed analysis on the charge distribution of thermoelectric materials has been reported in subsequent chapters.

Keywords

Rietveld Method, JANA2006, Structure, Refinement, Growth, Data Collection

Contents

2.1 INTRODUCTION

The discovery and production of highly efficient and cost effective novel materials for thermoelectric devices need in-depth knowledge of the structural details of the materials on a microscopic level. Material characterization is an important and difficult task in developing efficient thermoelectric materials, because, these materials need to be good electrical conductors and at the same time bad thermal conductors, *i.e.,* Phonon Glass Electron Crystal (PGEC) [Slack, 1995]. These materials need to allow freely the current carrying electrons, at the same time, restrict to the maximum extent, the flow of thermal electrons as well as phonons, *i.e.*, thermal effect, due to lattice vibrations. This is not an easy task unless one has complete knowledge of the structural information like lattice parameter, carrier concentration, type and nature of dopant, band gap energy, thermal Debye–Waller factor and the crystal symmetry.

2.2 X- RAY DIFFRACTION

The ideal method for solving crystal structures is single-crystal X-ray diffraction. Growing single crystals of appropriate size, however, is often difficult or even impossible, whereas powder samples are readily available for analysis by powder diffractometry. Despite the problems associated with peak overlap, a high-quality powder diffraction pattern generally contains enough information for unambiguously determining the corresponding crystal structure. A number of techniques are available to index the XRD pattern, which allow cell parameters to be derived from the positions of diffraction peaks. Knowledge of systematic absences can help to determine the most likely space groups. Taking into account the molecular connectivity and the well-known geometry of certain molecular fragments, the atomic arrangement in the unit cell can be described by a small number of parameters. It is possible to determine these structural parameters from the intensity distribution of the powder diffraction pattern, although solving crystal structures from powder diffraction data remains a difficult computational problem for more complex crystals.

2.3 RIETVELD METHOD

Rietveld technology [Rietveld, 1969; Young, 1993] is one of the well-known profile fitting technologies for the average structural analysis of crystalline materials, which uses powder samples. The structural parameters can be refined by considering a model structure and minimizing the weighted reliability factor, wR_p, by the least squares refinement process. Flexibility is provided through a wide range of refinement parameters like unit cell, atomic peak profile, asymmetry, crystallite size, strain broadening, preferred orientation, background, zero-point shift and Bragg intensities.

2.4 RIETVELD REFINEMENT USING JANA2006

JANA2006 [Petříček *et al.,* 2006] is (latest version of JANA2000) a crystallographic software, which allows the user to analyze the structural parameters like lattice parameters, unit cell volume, isotropic and anisotropic thermal vibration parameters and fractional positional coordinates of atoms in the unit cell. The relative proportions of elements in a mixed system can also be studied. The refinement is carried out by utilizing both the intensities of the Bragg peaks with diffraction angles as well as the background intensities. Legendre polynomials with up to 36 terms can be used to refine the background profile. Preferred orientation is often the greatest problem when trying to solve a structure. Although it is possible to evaluate the preferred orientation [Altomare *et al.*, 1994], this is still rare. Theoretically, one can treat the texture parameters like any other parameter in the optimization, using the March-Dollase model [March, 1932; Dollase, 1986] and this requires only 3 parameters: the [h k l] coordinates of the preferred orientation vector (only 2 independent parameters) and the March coefficient describing the type (plate or needle) and intensity of the preferred orientation. Asymmetry corrections can also be performed using Berar and Baldinozzi method [Baldinozzi and Berar, 1993]. The Pseudo-Voigt function [Thomson *et al.,* 1987], the summation of a Lorentzian and Gaussian function is used for profile refinements. In the present work, both JANA2006 [Petříček *et al.,* 2006] and the earlier version JANA2000 [Petříček *et al.,* 2000] were used for the analysis of the average structure of the thermoelectric materials considered in this book.

2.5 PRESENT WORK ON STRUCTURAL REFINEMENTS

In the present work, out of seven thermoelectric materials chosen for this study, the following five thermoelectric materials (1) Mg_2Si (2) PbTe (3) $Bi_{1-x}Sb_x$ with x=0.2 (4) Bi_2Te_3 and (5) Sb_2Te_3 have been studied for their structural analysis using JANA2006 [Petříček *et al.,* 2006] with the powder X-ray diffraction data sets. The powder X-ray diffraction data sets were collected in the 2θ range from 10° to 120° with step size 0.05° (except for $Bi_{1-x}Sb_x$, for which 2θ range is from 10° to 70°) at National Institute for Interdisciplinary Science and Technology (NIIST), CSIR, Trivandrum, India, using X'-PERT PRO (Philips, Netherlands) X-ray diffractometer with a monochromatic incident beam of a wavelength of 1.54056 Å, offering pure Cu-$K\alpha_1$ radiation.

The other two chosen thermoelectric materials (6) $Sn_{1-x}Ge_xTe$ with x=0.12 and 0.25 and (7) InSb were analyzed in a single crystal form using the single crystal X-ray diffraction data sets collected using the CAD-4 X-ray diffractometer with MoKα X-radiation (λ=0.7107Å) and graphite as the monochromator. The data set was collected with several psi-scan sets resulting in a transmission factor of about 1.

2.5.1 MAGNESIUM SILICIDE (Mg_2Si)

The literature survey on the material magnesium silicide (Mg_2Si) reveals that it has a face centered cubic lattice with an antifluorite structure with Si^{4-} ions occupying the corners and face center positions of the unit cell and Mg^{2+} ions occupying eight tetrahedral sites in the interior of the unit cell. There are also four interstitial sites [Madelung, 2004]. This structure follows the space group Fm-3m. There are four Mg_2Si molecules in the cubic unit cell with a cell constant of a=6.39 Å [Wyckoff, 1963]. Also, it is found that magnesium silicide (Mg_2Si) has proven to be a semiconductor with promising thermoelectric properties in the temperature range between 500K to 800K [Boriseneko, 2000; Morris, 1909; Redin, 1916; Kajikawa *et al.,* 1998; Labotz *et al.,* 1963; Noda *et al.,* 1992; Mahan *et al.,* 1996]. It is also a narrow band gap semiconductor with an indirect band gap of (0.6-0.8) eV [Morris *et al.,* 1958; Stella *et al.*, 1964]. This material is expected to be applicable in highly efficient solar cells, and has potential detector applications in the 1.2 mm to 1.8 mm infrared range, relevant for optical fibers [Leong *et al.,* 1997; Lange, 1997].

Compared with other thermoelectric materials operating in the same conversion temperature range such as PbTe and $CoSb_3$, important aspects of Mg_2Si include that it has been identified as an environment friendly material, as its constituent elements are abundant in the earth crust and it is nontoxic [Kondoh *et al.,* 2003; Krivosheeva *et al.,* 2002]. Because of these promising aspects, this material has been chosen for our structural analysis.

2.5.1.1 STRUCTURAL REFINEMENTS ON Mg_2Si

Powder X-ray diffraction studies were undertaken by using analar grade (99.99%) samples of Mg_2Si purchased from Alfa Acer, Johnson's Mathew Company. The X-ray intensity data have been collected from well grounded and sieved (400 mesh) samples of Mg_2Si. A standard software package, unit cell [Holland and Redfern, 1997] is used to refine the cell parameters of Mg_2Si using the observed 2θ values. The results of the unit cell refinement have been represented in table 2.1. The refined cubic cell parameter is found to be 6.3899(1) Å, whereas the reported [Wyckoff, 1963] value is 6.39 Å.

The average structural properties of the material were analyzed using the software package JANA 2000 [Petříček *et al,* 2000] (earlier version of JANA2006), which uses X-ray intensity data and works under the principle of Rietveld formalism [Rietveld, 1969]. The structure was refined by considering a basic structural model with a space group Fm-3m. The number of electrons in the unit cell $F_{(000)}$ is found to be 152. The observed and calculated intensity profiles along with the difference curve of Mg_2Si are shown in figure 2.1. The refined structure factors along with the standard deviation between the observed and calculated structure factors for various Bragg positions are given in table 2.2. The structural parameters of Mg_2Si obtained by the Rietveld refinement [Rietveld, 1969] are given in table 2.3.

Table 2.1 The observed and calculated d (inter planar distance) and 2θ values of Mg_2Si obtained from unit cell refinement

h k l	d_{obs}	d_{cal}	Δd	$2\theta_{obs}$	$2\theta_{Cal}$	Δ2θ
1 1 1	3.69329	3.67450	0.01879	24.076	24.201	-0.125
0 0 2	3.19627	3.18221	0.01406	27.890	28.016	-0.126
2 0 2	2.25407	2.25016	0.00391	39.965	40.037	-0.072
1 1 3	1.92089	1.91894	0.00195	47.282	47.333	-0.051
2 2 2	1.83887	1.83725	0.00162	49.529	49.575	-0.047
0 0 4	1.59120	1.59110	0.00009	57.906	57.909	-0.004
3 3 1	1.45978	1.46010	-0.00032	63.696	63.681	0.016
0 2 4	1.42309	1.42313	-0.00003	65.540	65.539	0.002
4 2 2	1.29835	1.29913	-0.00078	72.780	72.729	0.051
5 1 1	1.22395	1.22483	-0.00088	78.002	77.936	0.067
4 4 0	1.12510	1.12508	0.00002	86.413	86.415	-0.002
5 3 1	1.07551	1.07578	-0.00027	91.483	91.453	0.029

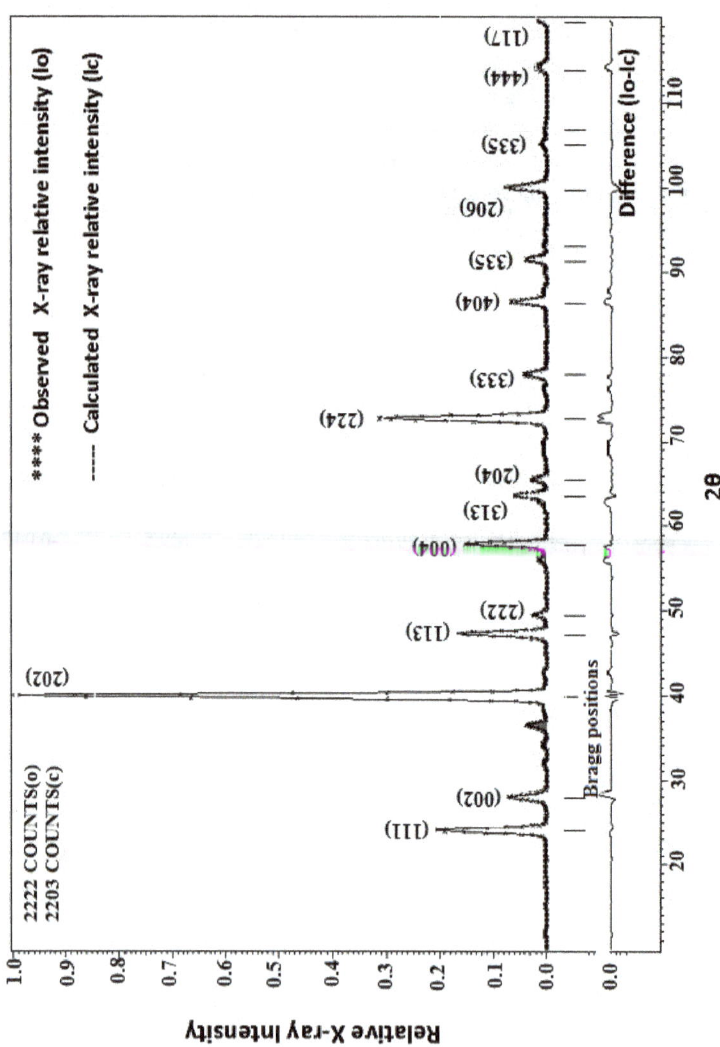

Figure 2.1 Relative X-ray powder profile of Mg_2Si

Table 2.2 The structure factors of Mg_2Si obtained using JANA 2000 software

h	k	l	F_o	F_c	$\sigma(F_o)$
1	1	1	44.46	44.45	0.528
0	0	2	32.27	30.93	0.697
2	0	2	97.25	98.38	0.536
1	1	3	32.99	32.71	0.465
2	2	2	21.99	23.49	0.886
0	0	4	79.69	78.43	1.139
3	1	3	25.88	27.38	0.623
2	0	4	16.56	17.25	0.446
1	1	5	23.63	23.53	0.442
3	3	3	23.66	23.53	0.347
4	0	4	54.43	53.53	1.191
3	1	5	20.57	20.38	0.626
4	2	4	7.31	8.91	0.935
0	0	6	7.31	8.91	0.934
3	3	5	16.32	17.72	1.072
2	2	6	5.56	6.35	1.057
4	4	4	39.25	37.86	1.664
5	1	5	15.42	15.45	0.999
1	1	7	15.42	15.45	0.999

2.5.1.2 DISCUSSION OF THE RESULTS

A perfect profile matching between observed and calculated intensities is clearly visible from figure 2.1, which signifies the correctness of the refinement. Also, it is found from table 2.1, that the maximum and minimum differences between calculated and observed structure factors (F_o-F_c) are very small. The maximum difference is 1.60 for the planes [424] and [006] and the minimum difference is 0.01 for the plane [111]. The minimum and maximum standard deviation values $\sigma(F_o)$ obtained are 0.347 for the plane [333] and 1.664 for the plane [444] respectively, which are shown in table 2.2. These small standard deviation values between the observed and calculated structure factors are yet another confirmation for the correctness of the refinements. The profile reliability index value (Rp%) is very low, showing the perfect fitting of the observed and calculated intensity profiles. Three small diffuse peaks corresponding to some unknown material was identified and blocked in the refinement process, which can be noted in figure 2.1.

The refined cell parameter of the unit cell (a=b=c) is 6.3671(14) Å, which is in close agreement with the reported value of 6.39 Å [Wyckoff, 1963]. From table 2.3, it is found that the thermal vibration parameters of the individual atoms are large for both the atoms Mg and Si, which signifies the use of Mg_2Si as a thermoelectric material. Moreover, the thermal vibration parameter of Mg atom, B_{Mg} is larger than that of the Si atom. Hence, we can conclude that the thermoelectric properties in Mg_2Si depend more on the Mg atom, (because of more scattering of phonons) than Si atom.

Table 2.3 The structural parameters of Mg_2Si obtained using JANA 2000 software

Parameter	Value
a (Å)	6.3671(14)
B_{Mg} (Å^2)	1.906(93)
B_{Si} (Å^2)	1.361(91)
R_{obs} (%)	3.41
R_P(%)	8.67
GOF	1.43

a is the cell parameter;
B_{Mg} is the Debye-Waller factor for Mg atom;
B_{Si} is the Debye-Waller factor for Si atom;
GOF is the goodness of fitting

2.5.21 LEAD TELLURIDE (PbTe)

Lead telluride (PbTe) is one of the best materials used in the construction of thermoelectric generators working at intermediate temperatures of 450K to 900K. The PbTe generators were developed for special application in space exploration in 1960's, and now constitute the power supply unit from gas combustion heat [Uemura and Nishida, 1988]. The preparation of PbTe material was made by melt-growth and hot pressing, which resulted in homogeneous materials. Investigations of lead chalcogenides doped with different impurities are stimulated by their wide applications in optoelectronics and thermoelectricity [Ravich *et al.*, 1970].

Results of the measurements of thermoelectric properties of thin semiconductor micro wires of $Pb_{1-x}Tl_xTe$ (x=0.001 to 0.02, d = 5 to 100 μm) at room temperature, where d is the thickness of the film, which were grown from solution melt by the filling of quartz capillary are presented [Zasavitsky *et al.*, 2005]. It is reported for the samples corresponding to chemical composition with concentration of thallium from 0.0025 to 0.005, there is an alternative change of sign of thermoelectric power observed. In pure samples and samples with thallium concentration more than 1%, the thermoelectric power is positive in the entire temperature range. Though this material has been used as thermoelectric material for many years in space applications, the literature survey reveals only limited information on the internal structure of this material.

2.5.2.1 STRUCTURAL REFINEMENTS ON PbTe

The cell parameters of PbTe were refined using a unit cell package [Holland and Redfern, 1997] with the help of observed 2θ values. The refined cell parameter is found to be 6.4508(61) Å whereas the reported [Wyckoff, 1963] value is 6.454 Å. The results of the cell refinement have been represented in table 2.4. The raw X-ray intensity data of PbTe was refined using the Rietveld [Rietveld, 1969] refinement technique. The cell parameters and other structural parameters were refined by this method, using the software JANA2000 [Petříček *et al.*, 2000]. The fitted profile and the positions of Bragg peaks for PbTe have been shown in figure 2.2. The refined structure factors with $\sigma(F_o)$ values are given in table 2.5. The refined structural parameters are given in table 2.6.

2.5.2.2 DISCUSSION OF THE RESULTS

The refined cell parameter of PbTe using the unit cell refinement is found to be 6.4508(61) Å. The difference in the cell constants between reported [Wyckoff, 1963] and that obtained by the unit cell refinement is 0.0032 Å. The cell constants obtained from JANA2000 [Petříček *et al.*, 2000] refinement is 6.4571(67) Å. The difference in the cell constants between reported [Wyckoff, 1963] and that obtained by the Rietveld refinement is 0.0031 Å. The fine profile matching between the observed and calculated intensities is shown in figure 2.2, the difference curve is almost a straight line, which shows the perfectness of the refinement. From table 2.5, it is found that the difference between the observed and calculated structure factors is very small. The thermal vibration parameters obtained from the Rietveld refinement [Rietveld, 1969] is shown for the individual atoms. Large Debye-Waller factor is reported for both the metallic atoms Pb and Te (table 2.6). This signifies the use of PbTe for high temperature thermoelectric applications. Further, it is found that the lattice thermal parameter for Pb is much greater than for Te. This further signifies the use of PbTe for high temperature thermoelectric applications.

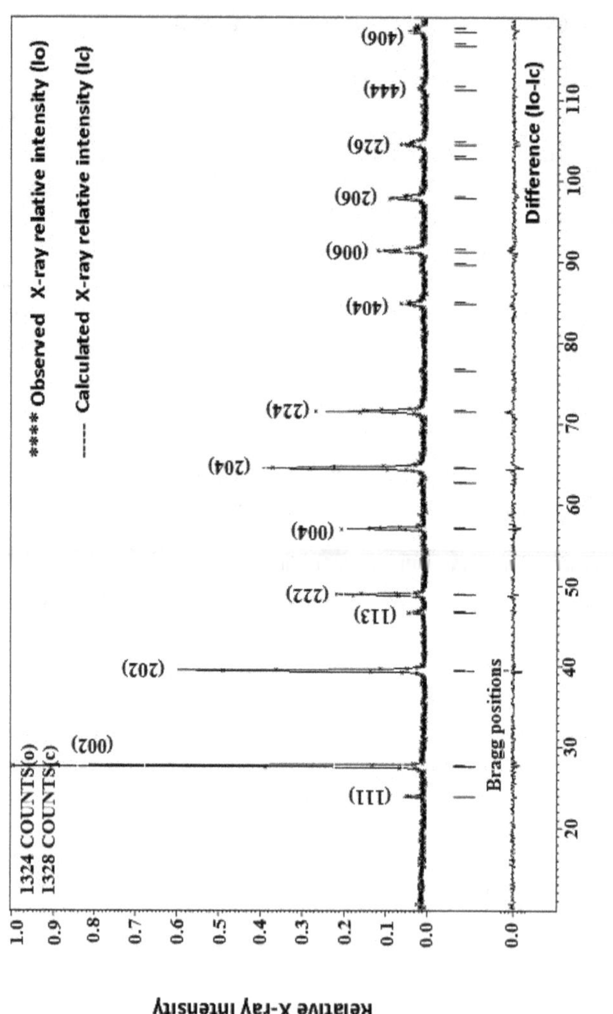

Figure 2.2 Refined X-ray powder profile of PbTe

Table 2.4 The observed and calculated inter planar distance (d) and 2θ values obtained from unit cell refinement of PbTe

h k l	d_{obs}	d_{cal}	Δd	$2\theta_{obs}$	$2\theta_{Cal}$	Δ2θ
0 0 2	3.2243	3.2251	-0.0008	27.643	27.636	0.007
2 0 2	2.2804	2.2785	0.0019	39.483	39.518	-0.036
1 1 3	1.9457	1.9469	-0.0012	46.642	46.610	0.031
2 2 2	1.8628	1.8635	-0.0007	48.850	48.831	0.019
0 0 4	1.6135	1.6120	0.0009	57.031	57.067	-0.036
2 0 4	1.4433	1.4452	-0.0018	64.509	64.418	0.091
2 2 4	1.3179	1.3182	-0.0003	71.529	71.510	0.019
3 3 3	1.2428	1.2423	0.0005	76.602	76.636	-0.034
0 4 4	1.1427	1.1414	0.0013	84.770	84.887	-0.117
2 4 4	1.0772	1.0777	-0.0005	91.301	91.246	0.056
2 0 6	1.0219	1.0221	-0.0001	97.826	97.815	0.011
2 2 6	0.9742	0.9735	0.0007	104.500	104.606	-0.106
4 4 4	0.9317	0.9318	-0.0001	111.527	111.522	0.005
0 4 6	0.8938	0.8941	-0.0004	119.049	118.971	0.078

Table 2.5 The structure factors of PbTe refined using JANA 2000 software

h	k	l	F_o	F_c	$\sigma(F_o)$
1	1	1	86.79	92.08	4.26
0	0	2	439.40	438.01	3.12
2	0	2	384.68	385.20	5.95
1	1	3	64.28	65.03	4.71
2	2	2	340.50	345.74	8.71
0	0	4	304.11	313.66	8.68
3	1	3	38.29	45.43	8.38
2	0	4	284.00	286.49	5.22
2	2	4	273.77	262.98	6.30
3	1	5	25.81	20.02	8.87
0	0	6	216.34	207.79	6.14
4	2	4	216.34	207.79	6.14
2	0	6	185.86	193.19	6.02
3	3	5	13.75	11.87	6.04
2	2	6	170.55	180.02	6.17
1	1	7	6.79	5.95	1.82
5	1	5	6.80	5.95	1.90
4	0	6	152.92	157.28	6.02

Table 2.6. The structural parameters of PbTe refined using JANA 2000 software

Parameter	Value
a (Å)	6.4571(0.0067)
B_{Pb} (Å^2)	2.410(0.202)
B_{Te} (Å^2)	0.989(0.186)
R_{obs} (%)	2.85

a is the cell parameter
B_{Pb} is the Debye-Waller factor for Pb atom
B_{Te} is the Debye-Waller factor for Te atom

2.5.3 BISMUTH DOPED WITH ANTIMONY ($Bi_{80}Sb_{20}$)

The bismuth (Bi) and antimony (Sb) based binary compounds have been reported as good thermoelectric materials with a currently achieved figure of merit (ZT) close to 2.2 at elevated temperatures [Hau *et al.*, 2004]. The thermoelectric figure of merit study of semiconducting alloy $Bi_{0.91}Sb_{0.09}$ thin films was reported by Sunglae *et al.*, 1998. Though numerous experimental works on this material are available in the literature, only limited information regarding the average and local structure is available. Hence, we have reported the growth and some of the structural based thermoelectric properties based on the Rietveld technique [Rietveld, 1969] using the software JANA 2000 [Petříček *et al.*, 2000].

2.5.3.1 GROWTH OF $Bi_{80}Sb_{20}$

The stochiometric compositions of analar grade (99.99%) bismuth (Bi) and antimony (Sb) purchased from Alfa Acer, Johnson's Mathew Company were used as the starting materials for the preparation of $Bi_{80}Sb_{20}$. The stochiometric compositions of fine powders of Bi and Sb were loaded in a thick walled quartz tube and evacuated to 10^{-6}torr using a high vacuum pump (Hind High Vacuum Co. Limited, Bangalore, India). The experimental setup used for quartz tube evacuation and sealing is shown in figure 2.3. The temperature of the furnace was raised slowly to 650°C. Then, after 20 hours of soaking time, the alloyed mixture was suddenly quenched on an aluminium plate kept at −5°C. The resulting poly crystals were finely ground, sieved and used for the powder X-ray diffraction data collection.

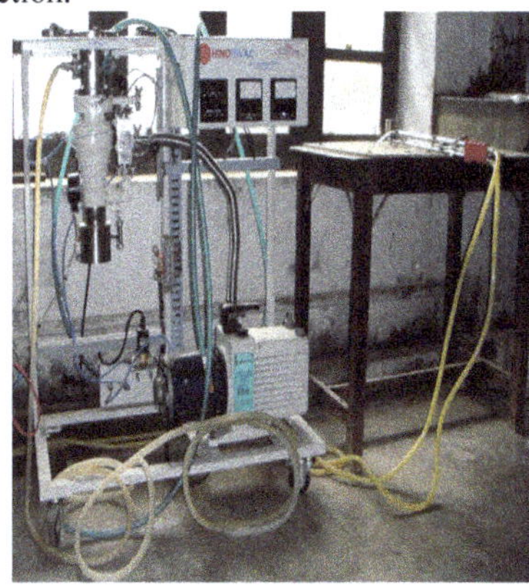

Figure 2.3 Experimental setup used for evacuation of quartz tube and sealing

2.5.3.2 STRUCTURAL REFINEMENTS OF $Bi_{80}Sb_{20}$

The cell parameters, thermal vibration parameters and the mass phase fractions of Bi, Sb and $Bi_{80}Sb_{20}$ were refined using the well-known powder profile fitting methodology Rietveld refinement [Rietveld, 1969] with the aid of the software package, JANA2006 [Petříček *et al.*, 2006], as done in the previous case. The fitted profiles and the positions of Bragg peaks for Bi, Sb and $Bi_{80}Sb_{20}$ are shown in figure 2.4, 2.5 & 2.6 respectively. In these figures, the dots represent the observed powder patterns and the continuous lines represent the calculated powder profiles of the respective samples. The difference of the observed profile and the fitted calculated profile is shown at the bottom of each figure. The small vertical lines in these figures indicate the positions of the Bragg peaks. In figure 2.6, the top vertical lines represent the Bragg positions of the $Bi_{80}Sb_{20}$ system, the middle ones represent those of Sb and the bottom lines represent the Bragg positions of Bi.

The refined structure factors for the samples Bi, Sb and $Bi_{80}Sb_{20}$ are given in the tables 2.7 to 2.9 respectively. In all three cases, there is a perfect matching between the observed and the calculated structure factors with very little difference between them, which proves the completeness of the refinement. The refined structural parameters are given in table 2.10.

2.5.3.3 DISCUSSION OF THE RESULTS

A perfect profile fitting with less deviation between observed and calculated intensities is a clear indication of the correctness of the refinement in all three cases Bi, Sb and $Bi_{80}Sb_{20}$, which are shown in figures 2.4, 2.5 & 2.6. The refined structure factors of $Bi_{80}Sb_{20}$ with σ(Fobs) values are given in table 2.9 showing good agreement of the observed and calculated structure factors with very low standard errors. The refined profile fittings further reveal nice matching of the observed and calculated profiles for all three systems.

The multiphase analysis of the $Bi_{80}Sb_{20}$ based on Rietveld technique [Rietveld, 1969] using JANA2006 [Petříček *et al.*, 2006] is shown in figure 2.6. It is found from the figure that the Bragg position of $Bi_{80}Sb_{20}$ exactly matches with the one of Bi. This is a clear indication of the perfect doping of Sb atoms in the atomic position of Bi. Also, the phase fractions obtained from the Rietveld refinements [Rietveld, 1969] show a mixed phase of Bi, Sb and $Bi_{80}Sb_{20}$. The multiphase analysis using the Rietveld refinements [Rietveld, 1969] reveal 83.91% of $Bi_{80}Sb_{20}$ formation and the remaining 16.09% was the parental elements of Bi and Sb, out of which 13.07% is Bi and 3.02% is Sb.

The comparative analysis of the background intensity of the parental elements and product element was carried out. It is found that the background intensity is highly suppressed in the resolved X-ray intensity profile of $Bi_{80}Sb_{20}$ compared to the parental elements Bi and Sb, which is shown in figure 2.6. This background suppression is found to be due to the high temperature growth, which is just like the effect due to the sintering process. The cell parameters obtained in this work by the refinements yielded consistent cell values. The difference in the cell constants between reported [Wyckoff, 1963] and that obtained in the refinements is 0.02054 Å and 0.0097 Å for a and c respectively for Bi. The same for Sb is 0.0043 Å and 0.0113 Å for a and c respectively. It is found that these cell values are highly comparable with those reported [Wyckoff, 1963]. The Debye-Waller factor for $Bi_{80}Sb_{20}$ is found to be 0.740347, which is much greater than for the individuals, Bi (0.24509) and Sb (0.151548). This may be one of the favorable conditions for the material $Bi_{80}Sb_{20}$ to be a good thermoelectric material.

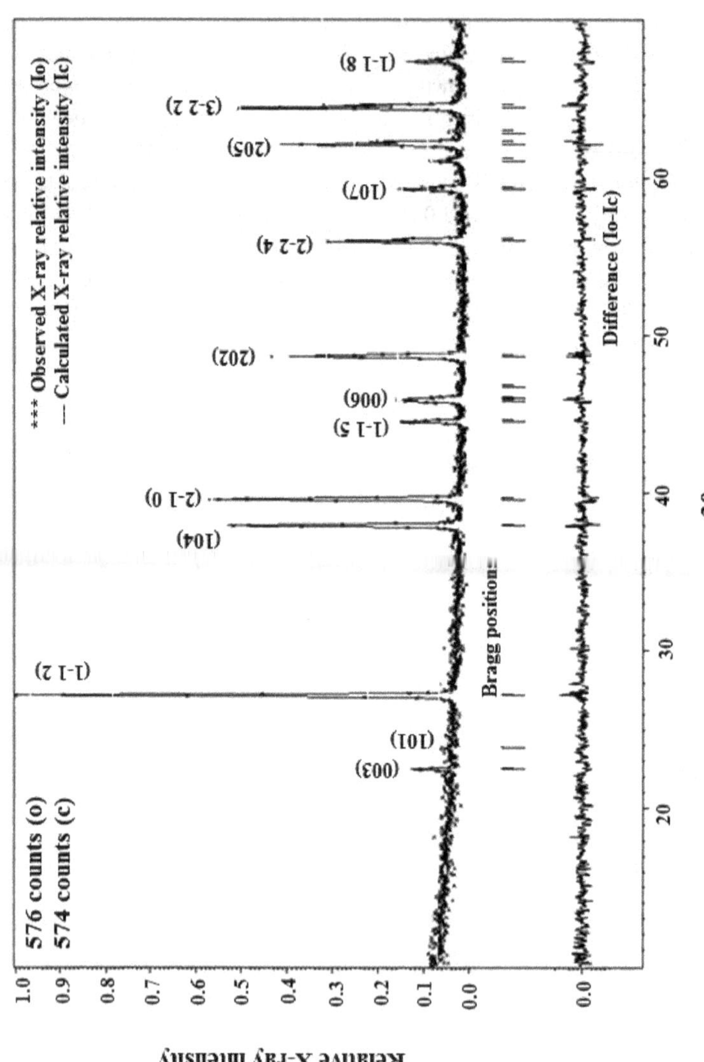

Figure 2.4 Refined X-ray powder profile of Bi

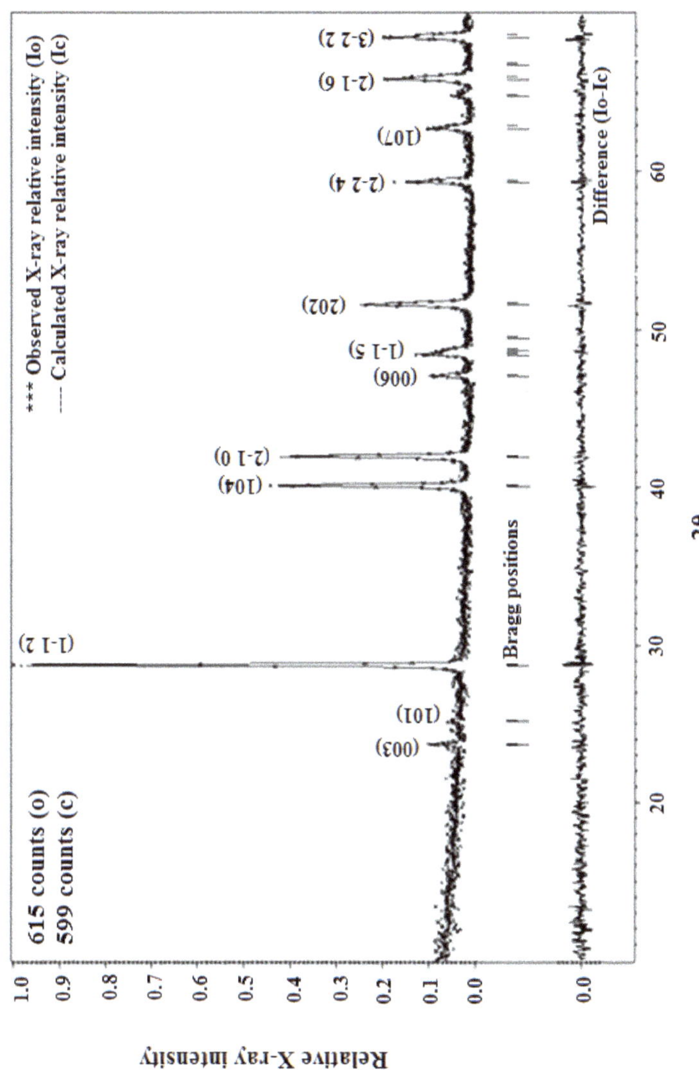

Figure 2.5 Refined X-ray powder profile of Sb

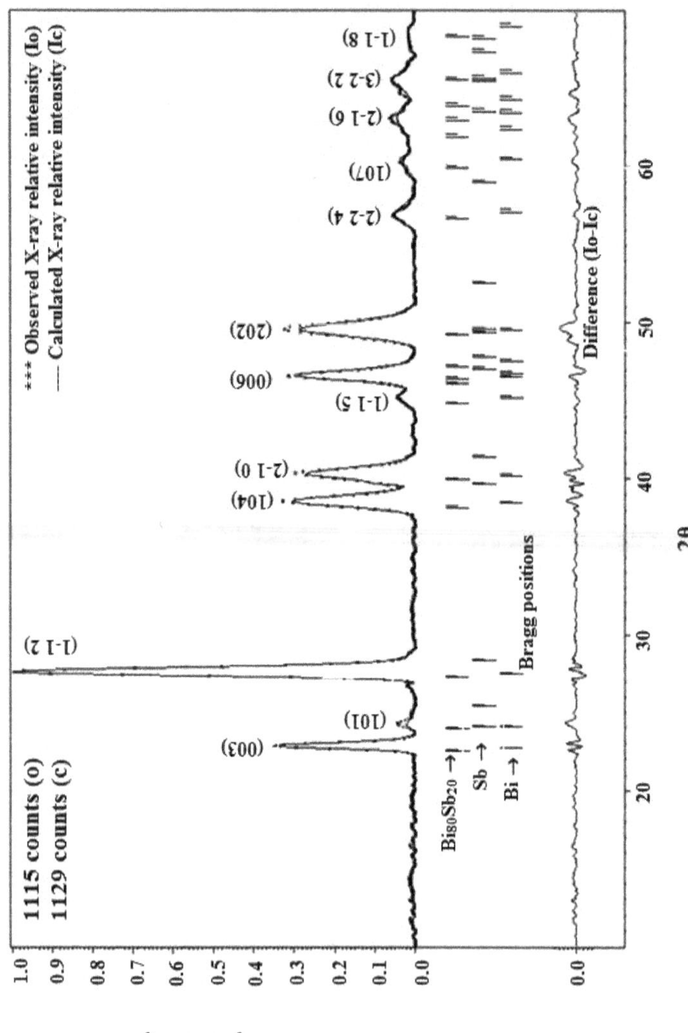

Figure 2.6 Refined Triple phase X-ray powder profile of Bi, Sb and $Bi_{80}Sb_{20}$

Table 2.7 The structure factors of Bi obtained using JANA 2006 software

h	k	l	Fo	Fc	$\sigma(F_o)$
0	0	3	148.19	151.65	11.2913
1	0	1	47.44	50.05	4.3755
1	-1	2	461.93	462.26	6.6743
1	0	4	430.87	436.22	5.9903
2	-1	0	497.64	504.36	6.8818
1	-1	5	265.38	258.44	6.0968
0	0	6	397.04	389.91	9.0116
2	-1	3	179.21	175.92	4.0193
2	-2	1	99.84	98.84	1.6100
2	0	2	521.03	525.49	11.0250
2	-2	4	482.49	477.21	13.0945
1	0	7	353.03	350.40	6.2568
2	0	5	274.86	276.95	4.3255
2	-1	6	440.00	442.01	5.6986
3	-1	1	24.57	24.69	0.3184
3	-2	2	587.04	591.42	7.6656
1	-1	8	379.91	381.26	5.4086

Table 2.8 The structure factors of Sb obtained using JANA 2006 software

h	k	l	Fo	Fc	$\sigma(F_o)$
0	0	3	82.87	83.54	5.8753
1	0	1	26.95	27.07	4.2842
1	-1	2	257.50	257.32	3.9234
1	0	4	227.26	229.00	3.8781
2	-1	0	229.88	230.10	3.7272
1	-1	5	115.87	119.35	3.2359
0	0	6	217.58	207.27	5.5052
2	-1	3	71.96	70.38	2.5206
2	-2	1	24.86	26.10	8.7629
2	0	2	220.67	210.41	5.6278
2	-2	4	178.08	190.49	5.8022
1	0	7	157.39	152.41	6.3033
2	0	5	99.98	104.48	6.5029
2	-1	6	173.32	174.32	4.5649
3	-1	1	18.55	18.05	1.3082
3	-2	2	181.58	180.44	4.5283

Table 2.9 The structure factors of $Bi_{80}Sb_{20}$ obtained using JANA 2006 software

h k l	Fo	Fc	σ(F$_o$)
0 0 3	83.99	83.82	0.9949
1 0 1	27.92	27.76	0.2810
1 -1 2	322.04	323.59	2.3659
1 0 4	272.07	270.33	2.1063
2 -1 0	278.58	278.65	1.9593
1 -1 5	102.40	103.04	0.6683
0 0 6	225.88	225.29	1.5457
2 -1 3	61.37	61.56	0.3565
2 -2 1	20.41	20.47	0.1486
2 0 2	248.67	241.08	2.2458
2 -2 4	212.08	206.86	0.2926
1 0 7	114.95	112.05	0.1592
2 0 5	79.94	79.81	0.2695
2 -1 6	175.53	174.85	0.5348
3 -1 1	15.98	15.91	0.0453
3 -2 2	188.75	187.92	0.4758
1 -1 8	145.34	144.62	0.4354

Table 2.10 The structural parameters of Bi, Sb and $Bi_{80}Sb_{20}$ obtained using JANA 2006 software

Parameter	Bi	Sb	$Bi_{80}Sb_{20}$
a(Å)	4.5538 (74)	4.3039 (23)	4.5395(43)
c(Å)	11.8832 (19)	11.2620 (60)	11.8980(12)
B(Å^2)	0.2450 (67)	0.1515 (82)	0.7403(66)
Position(z)(Å)	0.2330 (1)	0.2349 (5)	0.2263(1)
*w*R*p(%)*	14.49%	15.81%	9.27%

a &b are the cell parameters
B is the Debye-Waller factor

2.5.4 BISMUTH TELLURIDE (Bi_2Te_3)

Bismuth telluride is a narrow gap layered semiconductor with a trigonal unit cell. Bismuth telluride based polycrystalline materials are used for power generation or cooling applications [Doriane *et al.*, 2005]. Alloys of bismuth, antimony, tellurium, and selenium have been widely investigated in search of higher Seebeck coefficients around room temperature for power generation applications [Satterthwaite and Ure, 1957]. Recently, researchers have attempted to improve the efficiency of Bi_2Te_3 based materials by creating structures where one or more dimensions are reduced, such as nano wires or thin films. In one such instance, n-type bismuth telluride was shown to have an improved Seebeck coefficient of −287 μV/K at 54°C [Tan, 2005].

2.5.4.1 SAMPLE PREPARATION

High quality powder samples of Bi_2Te_3 and Sb_2Te_3 (Sb_2Te_3 Rietveld refinements will be discussed in consecutive chapters), purchased from Alfa Acer, Johnson Mathew Company, were kept inside a long quartz tube, which was sealed and annealed. Quartz tubes have the property to withstand temperatures of up to 1200°C. Selecting the quartz tube without any minute cracks and proper length is an important process, which helps in sealing the tubes later on. Initially, one end of the quartz tube was sealed, by means of a sealing gun and then loaded with the powder samples. After proper sealing, the tube was checked for any air leakage and then carefully placed inside the muffle furnace. While keeping the quartz tubes inside the furnace, proper care was given to protecting them from being cracked, due to thermal expansion. The samples of Bi_2Te_3 and Sb_2Te_3 were annealed to a temperature range of 500°C because both the samples have melting point of 583°C. The annealing process helps in removing the strain in the samples and even out the orientation of the atoms. The prepared samples are shown in figure 2.7.

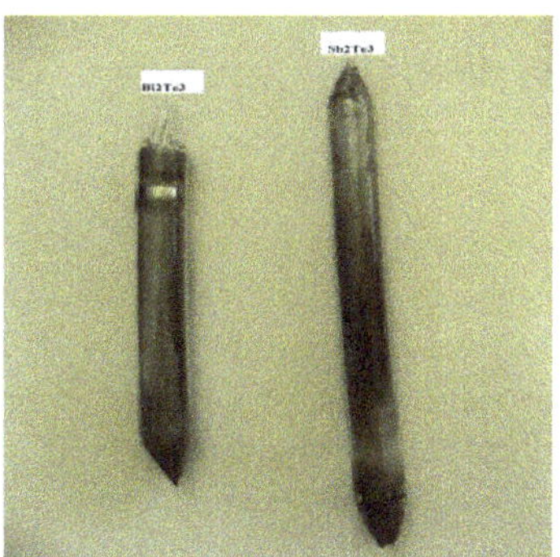

Figure 2.7 Annealed samples of Bi2Te3 and Sb2Te3 inside the quartz tube

2.5.4.2 STRUCTURAL REFINEMENTS ON Bi_2Te_3

X-ray powder data sets were collected for the prepared samples using monochromatic incident beam of Cu- Kα (1.54056 Å) radiation using X-PERT-PRO (Philips, Netherlands) X-ray diffractometer, with 2θ ranging from 10° to 120°. The Rietveld refinement [Rietveld, 1969] method was used for refining structural parameters like fractional coordinates, atomic displacement parameters, occupation factors and lattice parameters using software Jana2006 [Petříček *et al.*, 2006] directly from whole powder diffraction patterns. In this method, the observed profiles are matched with constructed profiles by using pseudo-Voigt [Wertheim *et al.*, 1974]. Thompson, Cox & Hastings [Thompson *et al.*, 1987] have modified the profile shape function to some extent that accommodates various Gaussian full width half maximum (FWHM) parameters and Scherer coefficient P for Gaussian broadening. The profile asymmetry is introduced by employing multi term Simpson rule integration devised by Howard [Howard, 1982] that incorporates symmetric profile shape function with different coefficients for weights and peak shift. JANA2006 [Petříček *et al.*, 2006] also employs the correction for preferred orientation which is independent of diffraction geometry using the March-Dollase function [March, 1932; Dollase, 1986]. The calculated profiles thus evolved are compared with the observed ones.

In refining the X-ray powder data, the space group was set as $R\bar{3}m$ with initial cell parameters of a=b=4.2641Å and c=30.4664Å. The Rietveld [Rietveld, 1969] refined profile is shown in figure 2.8. The refined structure factors for Bi_2Te_3 are tabulated in table 2.11. The refined structural parameters and the reliability indices are given in table 2.12.

2.5.4.3 DISCUSSION OF THE RESULTS

The refined cell constant values are found to be a=b=4.3935 (1) Å and c=30.5467 (8) Å, which are in very good agreement with the reported ones [Wyckoff, 1963]. The calculated error values are given within the parenthesis for that particular cell constant. The reliability index between the observed and calculated values is R_{obs}=3.66%. The error percentage for profile fitting between the observed and calculated ones is R_p=7.52%. The goodness of fit value is (GOF) 0.49. The intensity profile reveals nice matching between the observed and calculated intensities. The list of the structure factors corresponding to various Bragg positions are listed in table 2.11. The difference between the calculated and observed structure factors is found to be very low, therefore, indicating correctness of the refinement.

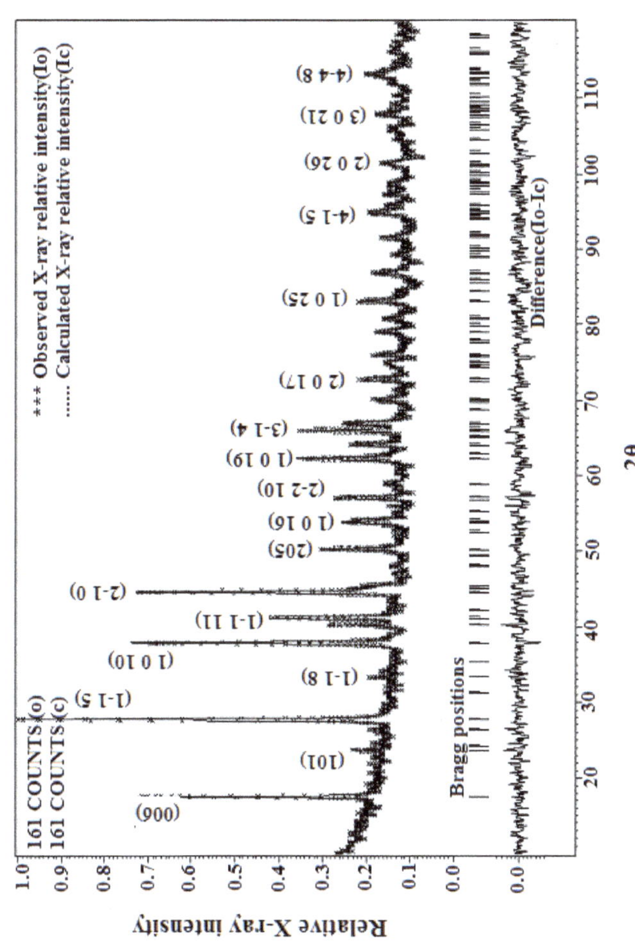

Figure 2.8 *Refined X-ray powder profile of Bi_2Te_3*

Table 2.11 The structure factor of Bi_2Te_3 obtained from JANA2006 software

h	k	l	F_{obs}	F_{cal}	$\sigma(F_{obs})$
0	0	6	228.32	234.89	5.92
1	-1	5	793.70	778.88	15.36
1	0	10	628.88	665.26	5.12
1	-1	11	303.19	280.09	15.91
2	-1	0	659.93	686.48	19.05
0	0	15	568.92	569.15	11.58
2	-1	6	215.36	214.71	4.85
1	0	5	215.45	202.46	10.97
2	0	5	615.66	610.55	22.67
1	0	16	290.42	304.57	11.46
2	0	8	109.00	114.26	4.53
0	0	18	238.43	250.02	9.17
2	-2	10	540.31	538.93	25.07
2	0	11	304.44	252.82	41.6
2	-1	15	447.58	469.67	14.04
1	0	19	47.44	48.96	2.12
2	-2	5	160.13	161.53	8.7
0	0	21	322.89	311.35	18.66
3	-1	1	149.74	153.14	20.75
3	-2	2	6.39	6.13	0.6
1	-1	20	435.90	403.60	14.03
3	-1	4	118.62	111.30	3.66
3	-2	5	501.1	507.84	19.49
2	-2	16	273.87	272.99	19.59
3	-2	8	92.70	92.27	6.48
2	-1	18	205.39	204.11	14.35
2	0	17	22.22	22.99	1.28
1	0	22	3.84	3.93	0.21
3	-1	10	445.74	454.53	22.71
3	-2	11	237.00	231.11	20.31
0	0	24	73.74	72.03	5.68
3	0	0	488.97	472.74	34.07
3	-3	3	41.80	40.16	2.49
3	0	3	41.80	40.16	2.49
1	-1	23	246.67	239.66	18.51
2	-1	21	260.66	278.45	17.82
2	0	20	351.21	346.73	32.22
2	0	29	105.77	105.62	5.9
4	-1	17	14.47	15.43	1.67
3	-2	26	207.33	222.23	24.2
4	-2	3	33.94	34.16	4.22
3	0	15	346.93	347.24	27.25
3	-3	15	346.93	347.24	27.25
4	-2	6	176.46	177.27	5.88
3	-1	19	74.49	74.97	5.97
4	-3	1	150.70	144.18	8.56
4	-1	2	4.42	4.36	0.22
1	0	28	228.83	227.01	11.12
4	-3	4	115.48	115.46	5.08
4	-2	9	93.25	93.19	4.12
4	-1	5	367.77	376.71	18.53
2	-2	25	286.17	262.03	38.12
4	-3	7	15.77	14.91	1.83
0	0	30	214.55	224.71	23.26
4	-1	8	63.95	66.76	6.67
3	0	18	59.57	146.82	14.9
3	-3	18	59.57	146.82	14.9
1	-1	29	98.86	104.35	10.66
2	0	26	264.51	244.99	20.07
3	-1	22	6.24	5.82	0.46
4	-3	10	367.28	342.42	27.07
4	-2	15	303.04	306.15	37.96
4	-3	5	93.20	98.76	12.24
3	0	21	220.57	227.91	14.86
3	-3	21	220.57	227.91	14.86
1	0	31	224.96	230.32	14.89
4	0	1	56.51	58.50	8.42
4	-1	14	83.48	84.26	5.17
4	-4	2	4.89	4.96	0.3
2	-2	28	215.86	196.32	27.72
4	0	4	117.41	114.33	12.47
4	-4	5	356.77	331.90	48.67
3	-1	25	212.30	234.62	27.61
4	0	7	11.78	11.81	0.73
1	-1	32	34.05	33.43	2.03
0	0	33	214.75	210.80	12.8
2	-1	30	204.03	202.95	11.11
4	-3	16	206.12	205.95	10.88
4	-4	8	59.26	59.04	3.09
4	-2	18	128.71	127.94	6.93

Table 2.12 The structural parameters of Bi_2Te_3 obtained using JANA 2006 software

Parameter	Values
a (Å)	4.3935(1)
b (Å)	4.3935(1)
c (Å)	30.5467(8)
Cell Volume($Å^3$)	510.65(3)
Density (gm/cc)	7.8090(4)
R_{obs}(%)	3.66
wR_{obs} (%)	4.14
R_p(%)	7.52
wR_p(%)	9.9
GOF	0.49

a, b & c are the cell parameters
GOF is goodness of fitting

2.5.5 ANTIMONY TELLURIDE (Sb_2Te_3)

Antimony Telluride (Sb_2Te_3) is a narrow band gap semiconductor with a homologous layered crystal structure. Both Sb_2Te_3 and Bi_2Te_3 are semi-metal alloys with good electrical conductivity (σ) and low thermal conductivity (k) [Zou *et al.*, 2001; Kim *et al.*, 2002]. They can be used for many different applications which typically fall into two general categories *ie.,* power generation and cooling devices [Doriane *et al.*, 2005]. Sb_2Te_3 in single crystal form has attracted great attention for high electrical conductivity, of the order of (3.13–5.26) x 10^5 $(\Omega/m)^{-1}$ at 300 K and low thermal conductivity of the order of (1.6–5.6) W/m/K at 300K [Rowe, 2006]. Unfortunately, it was reported that the Seebeck coefficient of the undoped Sb_2Te_3 single crystal is too low and of the order of (83–92) μV/K at 300K [Rowe, 2006]. However, it is reported that higher S can be attained by introducing nanostructures, for example, 125 μV/ K at room temperature was reported for a film of Sb_2Te_3 nano plates [Shi *et al.*, 2008] and 185 μV/K at 505 K for a thin film grown by co-evaporation [Zou *et al.*, 2001]. Unfortunately, the measured electrical conductivity of both films were only 10^2 $(\Omega/m)^{-1}$, which was far lower than that of the undoped Sb_2Te_3 single crystal. Though many reports on this thermoelectric behavior are available in the literature, only limited information is available regarding the structure, electron density distribution and bond length distribution. Since the internal structure has great impact on the thermoelectric behavior, an attempt was made to study the electron level properties in this work.

2.5.5.1 STRUCTURAL REFINEMENTS ON Sb_2Te_3

The structural parameters of Sb_2Te_3 were refined using the well-known profile fitting methodology, called the Rietveld method [Rietveld, 1969]. X-ray powder data sets were collected for the prepared samples using a monochromatic incident beam of Cu-Kα (1.54056 Å) radiation using a X-PERT-PRO (Philips, Netherlands) X-ray diffractometer, with 2θ ranging from 10° to 120°. The fractional coordinates, atomic displacement parameters, occupation factors and lattice parameters were calculated using the software JANA2006 [Petříček *et al.,* 2000 and updated in 2006], which works under the principle of the Rietveld method [Rietveld, 1969]. The experimental whole powder diffraction pattern was matched perfectly with the calculated one, by refining various parameters, which are available in the JANA2006 [Petříček *et al.,* 2006] software.

The initial cell parameters were chosen as a=b=4.2641Å, c=30.4664Å, α=β=90°; γ=120° with space $R\bar{3}m$. After refining the structural and profile parameters, the refined profiles show a very good agreement with calculated and observed patterns, as shown in figure 2.9. The calculated intensity profile is matched along with the observed intensity profile. Each peak was representing a single plane of (h k l). The error profile shows the difference between the calculated and observed peaks. The refined structure factors were tabulated in table 2.13. The refined structural parameters and the reliability indices are given in table 2.14.

2.5.5.2 DISCUSSION OF THE RESULTS

The refined cell constant values are found to be a = b = 4.2675(4) Å, c = 30.4635(3) Å, which is in close agreement with the reported cell values [Wyckoff, 1963]. The values in the parenthesis are the calculated error values for that particular cell constant. The reliability index between observed and calculated values is R_{obs} = 8.28%. The error percentage for profile fitting between the observed and calculated one is R_p = 4.69%. The correctness of the fit value (GOF) is 0.34.

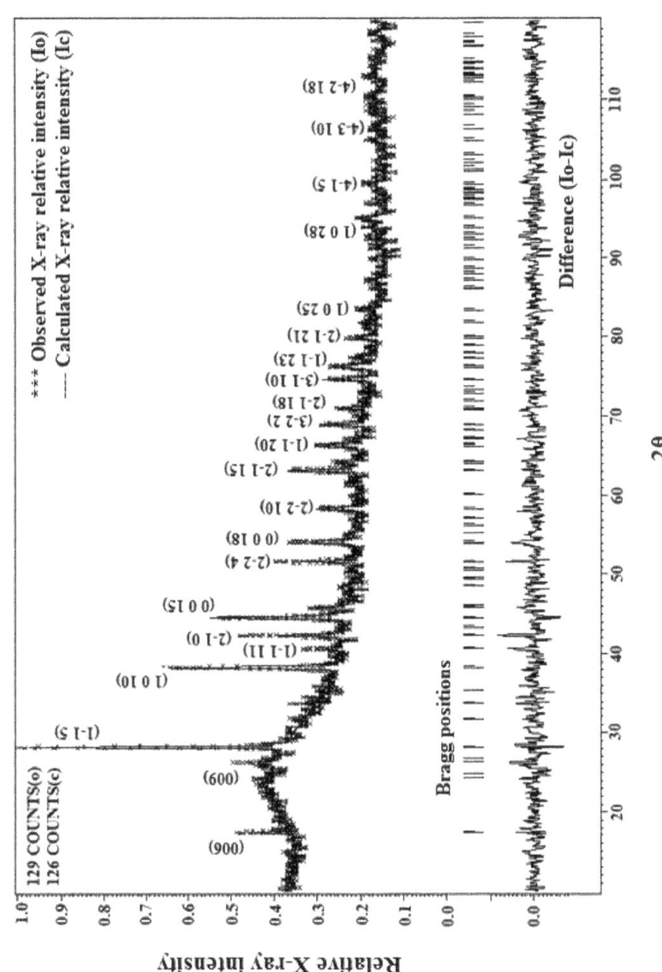

Figure 2.9 *Refined X-ray powder profile of* Sb_2Te_3

Table 2.13 The structure factors of Sb_2Te_3 obtained from JANA2006 software

h	k	l	Fo	Fc	σ (Fo)
0	0	6	85.61	83.44	13.0625
1	-1	5	630.63	642.04	19.757
1	0	7	100.70	77.03	36.6792
1	-1	8	154.29	130.79	45.5415
1	0	10	547.19	549.96	18.3603
1	-1	11	144.41	125.53	67.5704
2	-1	0	663.31	579.31	38.4097
2	-1	3	36.79	41.38	13.9752
0	0	15	430.24	465.09	15.7736
1	0	13	253.92	204.42	22.409
2	-1	6	85.32	66.38	7.2997
1	-1	14	146.40	109.58	25.3796
2	-2	1	17.45	10.03	7.1295
2	0	2	49.52	21.70	20.5095
2	-1	9	77.19	81.39	30.8649
2	-2	4	35.71	40.54	14.9369
2	0	5	582.33	523.22	41.9074
1	0	16	175.75	152.90	12.6017
0	0	18	312.60	272.37	22.3812
2	0	8	155.46	109.62	30.1082
2	-1	12	61.98	82.10	32.2365
1	-1	17	73.65	82.96	30.8142
2	-2	10	441.83	463.17	39.8675
2	0	11	135.37	106.46	24.9739
2	-1	15	397.90	398.00	21.2008
1	0	19	103.24	103.33	5.4721
2	-2	13	209.72	175.85	22.3929
0	0	21	193.98	165.06	23.9752
2	0	14	84.99	93.28	5.2028
1	-1	20	297.93	329.32	20.8179
3	-1	1	6.94	9.23	2.1003
3	-2	2	27.55	17.77	7.7513
3	-1	4	34.39	34.31	33.1408
3	-2	5	439.92	453.66	22.4233
3	-1	7	61.43	53.207	4.8892
2	-2	16	134.97	132.87	17.1333
2	-1	18	240.12	237.49	30.5833
3	-2	8	151.38	96.30	34.2626

h	k	l	Fo	Fc	σ (Fo)
2	0	24	79.22	84.36	4.9262
3	-2	11	101.18	93.52	4.0835
1	-1	23	307.95	291.77	24.8649
3	0	0	441.07	433.58	60.0564
3	0	3	23.85	32.73	11.2357
3	-3	3	23.85	32.73	11.2357
2	-2	19	113.72	89.96	21.0696
3	-1	13	174.59	155.77	25.4365
2	-1	21	164.33	145.33	23.5506
3	-3	6	57.94	50.85	9.7601
3	0	6	57.94	50.85	9.7601
3	-2	14	69.23	81.46	7.8049
2	0	20	243.91	292.06	35.0058
3	0	9	37.31	60.58	8.4657
3	-3	9	37.31	60.58	8.4657
1	0	25	168.91	231.42	37.87
0	0	27	30.17	27.20	4.6708
3	-1	16	130.62	118.07	19.9103
1	-1	26	156.80	143.71	36.7485
2	-2	22	81.66	59.02	13.4618
3	-2	17	54.17	63.38	38.6225
2	-1	24	115.43	74.11	47.9751
2	0	23	242.17	261.19	50.6581
4	-2	0	360.91	388.30	121.828
4	-2	3	56.68	30.12	11.2375
3	0	15	293.14	315.29	28.5433
3	-3	15	293.14	315.29	28.5433
3	-1	19	74.28	79.39	7.3805
4	-2	6	45.74	46.05	2.8413
1	0	28	295.19	301.93	27.0631
4	-1	5	368.95	363.05	52.7831
1	-1	29	50.73	48.56	6.5823
4	-3	10	330.49	328.49	48.1152
4	0	1	8.60	8.52	1.4013
0	0	33	301.91	302.23	45.1329
4	-4	2	10.69	10.56	1.565
1	-1	32	4.31	4.31	0.5977
3	-1	25	174.15	191.12	26.1938

Table 2.14 The structural parameters of Sb_2Te_3 obtained using JANA 2006 software

Parameter	Values
a (Å)	4.2675 (4)
b (Å)	4.2675 (4)
c (Å)	30.4635(3)
Cell Volume(Å^3)	480.4525(1)
Density (gm/cc)	6.4918(14)
R_{obs}(%)	8.28
wR_{obs} (%)	12.15
R_p(%)	4.69
wR_p(%)	6.17
GOF	0.34

a, b & c are the cell parameters
GOF is goodness of fitting

2.5.6 $Sn_{1-x}Ge_xTe$ SINGLE CRYSTALS

Tin telluride is one of the prospective materials for thermoelectric applications. It crystallizes in the structure like NaCl. It is reported to have a band gap of 0.26 eV at 300K [Abrikosov *et al.*, 1975]. Several reports are available stating the possible usage of Tin Telluride doped with different materials for thermoelectric applications. But the structural details due to doping of germanium in tin telluride are nowhere found in the current literature.

2.5.6.1 GROWTH OF $Sn_{1-x}Ge_xTe$ SINGLE CRYSTALS

The single crystals of $Sn_{0.88}Ge_{0.12}Te$ and $Sn_{0.75}Ge_{0.25}Te$ were grown by the Bridgman technique using high purity (99.999% pure) starting materials Sn, Ge and Te in suitable weight proportions. Appropriate weights of the elements were loaded into the quartz ampoule, which was then evacuated to a pressure of approximately 10^{-6} Torr using a high vacuum pump. The evacuated quartz ampoule was then sealed. The quartz ampoule was placed in a vertical furnace whose temperature was increased slowly and maintained at about a temperature of 100 K above the melting point. The growth velocity and the temperature gradient for getting good single crystal are 0.2 mm per hour (moving from the hot zone to the cold zone of the furnace) and 100 K per cm. In order to reduce the precipitation of free Te, proper care was taken with (Ge + Sn): Te as stoichiometric. To produce samples in the desired compositional range, melts containing about 20–40% GeTe were used.

Single crystalline sections were selected from the tip regions of the boules, and cut into samples of about 2×3×1 mm size. Small cleaves of bulk from $Sn_{0.88}Ge_{0.12}Te$ and $Sn_{0.75}Ge_{0.25}Te$ were cut and made into fine spheres using a homemade crystal-spherizer. The strained surfaces of the spheres were removed using a suitable etching solution. After recording several X-ray photographs to check the quality of these samples, good quality single crystal spheres of $Sn_{0.88}Ge_{0.12}Te$ and $Sn_{0.75}Ge_{0.25}Te$ with radii 0.155(3) mm and 0.161(2) mm respectively were used for the X-ray data collection. Then the samples were dipped in liquid nitrogen to reduce the effect of extinction of X-rays. The X-ray diffraction data were collected using CAD-4, X-ray diffractometer with Mo-Kα X-radiation (λ=0.7071Å) and graphite as the monochromator. The cell refinement has been done within a θ range of 20°. The photograph of the single crystal sphere of $Sn_{0.88}Ge_{0.12}Te$ with radii 0.155(3) mm is shown in figure 2.10.

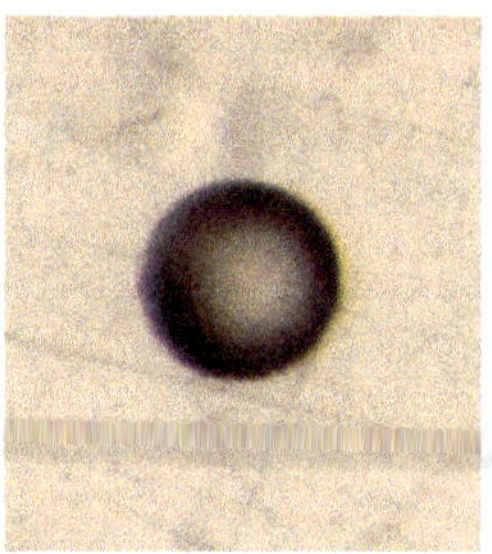

Figure 2.10 Single crystal sphere of Sn0.88Ge0.12Te with radii 0.155(3) mm

Table 2.15 The structure factor of Sn0.88Ge0.12Te obtained from JANA2006 software

h	k	l	F_{obs}	F_{cal}	$\sigma(F_{obs})$
1	1	1	30.29	22.69	3.3936
0	0	2	345.10	372.97	3.9738
0	2	2	369.04	340.62	3.7843
2	2	2	329.75	315.17	3.5584
1	1	3	25.84	2.10	2.4897
1	3	3	26.76	10.83	3.0275
3	3	3	28.34	16.78	6.0424
0	0	4	283.17	292.52	3.8811
0	2	4	269.95	271.94	3.1404
2	2	4	250.96	253.06	3.4440
0	4	4	220.16	219.78	3.3348
2	4	4	203.21	205.14	3.2751
4	4	4	162.27	168.09	5.0447
1	1	5	26.18	16.78	2.9522
1	3	5	25.70	21.07	2.3653
3	3	5	25.48	24.14	3.8890
1	5	5	23.77	26.29	3.7085
3	5	5	22.34	27.72	4.4500
5	5	5	20.68	29.01	8.4494
0	0	6	201.66	205.14	3.7294
0	2	6	189.28	191.70	2.6530
2	2	6	176.71	179.38	3.2189
0	4	6	156.56	157.73	2.5255
2	4	6	146.36	148.21	2.4912
4	4	6	121.46	123.96	3.2511
0	6	6	117.89	117.09	3.2811
2	6	6	113.43	110.73	3.3168
4	6	6	94.00	94.22	3.4203
6	6	6	73.83	73.21	6.1352

h	k	l	F_{obs}	F_{cal}	$\sigma(F_{obs})$
1	1	7	24.33	26.29	3.2293
1	3	7	22.58	27.72	2.8008
3	3	7	21.89	28.59	4.8421
1	5	7	20.94	29.01	3.0485
3	5	7	19.83	29.08	3.6481
5	5	7	18.75	28.47	5.5586
1	7	7	18.50	28.47	4.9881
3	7	7	17.24	27.88	6.1219
5	7	7	15.57	26.36	6.8119
0	0	8	131.30	131.40	5.2460
0	2	8	123.80	123.96	3.8784
2	2	8	119.06	117.09	5.4356
0	4	8	110.72	104.82	3.9733
2	4	8	107.49	99.33	4.0207
4	4	8	88.86	84.98	5.9043
0	6	8	83.45	80.81	4.2386
2	6	8	82.65	76.89	4.2152
4	6	8	72.29	66.49	4.3856
1	1	9	20.14	29.08	6.4897
1	3	9	20.37	28.88	5.1742
3	3	9	17.99	28.47	10.6476
1	5	9	17.56	27.88	6.3439
3	5	9	18.55	27.17	7.0893
0	0	10	85.49	80.81	5.8647
0	2	10	80.11	76.89	4.2528
2	2	10	78.79	73.21	6.0359
0	4	10	71.04	66.49	4.4028
2	4	10	68.65	63.42	4.5368
1	1	11	16.73	26.36	8.260

2.5.6.2 STRUCTURAL REFINEMENTS ON $Sn_{1-x}Ge_xTe$

The standard least-squares procedure using JANA 2006 [Petříček *et al.,* 2006] was adopted for refining the physical parameters including secondary extinction. All the necessary corrections were applied precisely and the refinement was carried out. The structure factors obtained from the Rietveld refinement [Rietveld, 1969] are tabulated in table 2.15 and 2.16 respectively for the systems $Sn_{0.88}Ge_{0.12}Te$ and $Sn_{0.75}Ge_{0.25}Te$. The results of the Rietveld refinement [Rietveld, 1969] for both the samples are given in table 2.17.

2.5.6.3 DISCUSSION OF THE RESULTS

The structure refinement using JANA 2006 [Petříček et al., 2006] gives reasonable values of the Debye-Waller factors for Sn(Ge) and Te in both Sn0.88Ge0.12Te and Sn0.75Ge0.25Te. From table 2.17, it is found that the reliability indices arc very low indicating the correctness of the refinements. The Debye-Waller factors of both the atoms Sn and Te increase as the concentration of germanium increases. This may be due to the increased lattice distortion arising out of the doping process. This suggests that the optimum doping of germanium results in a higher Debye–Waller factor, which favorably increase the figure of merit in germanium doped Sn(Ge)-Te thermoelectric materials.

2.5.7 INDIUM ANTIMONIDE (InSb) SINGLE CRYSTAL

Indium Antimonide (InSb) material is a binary (V-III) semiconductor which crystallizes in the zinc-blend structure [Udayashankar and Bhat, 2001]. Previous studies of InSb semiconductors, showed that the optical band gap 0.18 eV at 20°C is very close to the optical band of atmosphere window at (3-5μm) region [Morten and King, 1965]. The InSb devices as thermal cameras are important mostly in thermal detection of hot bodies at temperature range of (400-700) °C. It is the material of choice for magnetic-field sensing devices such as Hall sensors and magneto resistors [Heremans et al., 1993], speed-sensitive sensors [Hyun et al., 2007], millimeter wave devices [Senthilkumar et al., 2005], and magnetic sensors [Miyazaki and Adachi, 1991]. Further, the experimental results have shown that bulk InSb is a promising candidate for thermoelectric power generation with figure of merit 0.6 at a temperature of 673K [Yamaguchi et al., 2005].

Table 2.16 The structure factor of $Sn_{0.75}Ge_{0.25}Te$ obtained from JANA2006 software

h	k	l	F_{obs}	F_{cal}	$\sigma(F_{obs})$
1	1	1	50.25	44.87	1.5173
0	0	2	422.42	421.40	4.5534
0	2	2	397.19	390.51	4.0076
2	2	2	356.80	364.25	4.1829
1	1	3	44.28	41.50	1.2107
1	3	3	41.49	41.54	1.5448
3	3	3	40.17	41.01	3.3388
0	0	4	349.02	340.37	3.8547
0	2	4	313.42	318.15	3.3175
2	2	4	304.17	297.33	3.3751
0	4	4	243.34	259.55	2.8847
2	4	4	232.18	242.54	2.8137
4	4	4	191.82	198.48	3.5213
1	1	5	42.56	41.01	1.4185
1	3	5	38.36	39.78	1.2155
3	3	5	37.86	38.15	2.0696
1	5	5	33.56	36.36	1.8919
0	0	6	252.64	242.54	2.9783
0	2	6	229.39	226.72	2.5635
2	2	6	218.91	212.06	2.7470
0	4	6	179.25	185.92	2.1665
1	1	7	37.12	36.36	1.6058
1	3	7	33.56	34.55	2.2036
3	3	7	30.60	32.77	4.1588
0	0	8	156.94	153.63	3.5109
0	2	8	146.08	144.43	2.6733
2	2	8	138.10	135.89	3.5304
1	1	9	28.37	29.41	3.0284
1	3	9	25.77	27.83	2.6899
3	3	9	24.64	26.32	4.9855
0	0	10	98.28	90.61	3.7001
0	2	10	90.48	85.72	2.6835
2	2	10	86.43	81.14	3.7475
1	1	11	22.02	22.12	3.7662

Table 2.17 The structural parameters of $Sn_{0.75}Ge_{0.25}Te$ and $Sn_{0.88}Ge_{0.12}Te$ obtained using JANA 2006 software

Parameter	$Sn_{88}Ge_{12}Te$	$Sn_{75}Ge_{25}Te$
a (Å)	6.2713(4)	6.2158(5)
Space group	Fm-3m	Fm-3m
$B_{Sn,Ge}$ (Å^2)	1.83(8)	2.07(8)
B_{Te} (Å^2)	1.22(8)	1.42(9)
wR_p	4.33	3.40
GOF	1.71	2.11

a is the cell parameter
B is the Debye-Waller factor

2.5.7.1 DATA COLLECTION

Small spheres of InSb crystals were prepared from bulk crystals using a home-made crystal-spherizer and the strained surface was etched with a suitable solution so as to attain perfect spherical nature with little tolerance. A good quality single crystal sphere as shown in figure 2.11, with radius 0.0605(10) mm was chosen for structural analysis and single crystal X-ray diffraction data were collected using CAD-4 X-ray diffractometer with Mo-Kα X-ray radiation and graphite as the monochromator. The data set was collected with several psi- scan sets resulting in the transmission factor of approximately 1. Three standard reflections were monitored for every two hours (0.99 < decay < 1.0056) and a total of 867 reflections were measured. The quasi-forbidden reflections of the type h+k+l =4n+2 were also measured. Among these 867 reflections, 257 reflections are h+k+l =4n+2 type. The cell refinement has been done within a θ range of 6.00° to 16.96°. InSb belongs to a cubic structure with space group F43m (Zinc blende structure). The unit cell consists of four molecules at special positions 4(a) and 4(c).

2.5.7.2 REFINMENTS

The raw intensity data set of single crystal sphere of InSb was corrected for absorption. The μR value turned out to be 1.1. The absorption corrected data set was refined using the software package JANA 2006 [Petříček et al., 2006]. Table 2.15 represents the refined structure factors with σ(Fo) values after the final refinements. The reliability indices of these refinements are very low (Robs = 2.92% and wRobs = 4.99%). The allowed reflections for InSb are h+k+l=4n, h+k+l=4n±1 and h+k+l=4n+2, where n is integer. Among these type of reflections, the h+k+l=4n+2 type reflections are very weak and are called quasi-forbidden reflections. The theoretical structure factor expression for the quasi-forbidden, h+k+l=4n+2 type reflections is given by Fc= 4(fSb-fIn) where fSb and fIn are the atomic scattering factors of antimony and indium atoms respectively. Since these structure factors involve the difference in the scattering factors of constituent atoms, the expected experimental structure factors will be weak. The refined cell constants were found to be 6.463(6) Å, whereas the reported value is 6.479 Å [Wyckoff, 1963].

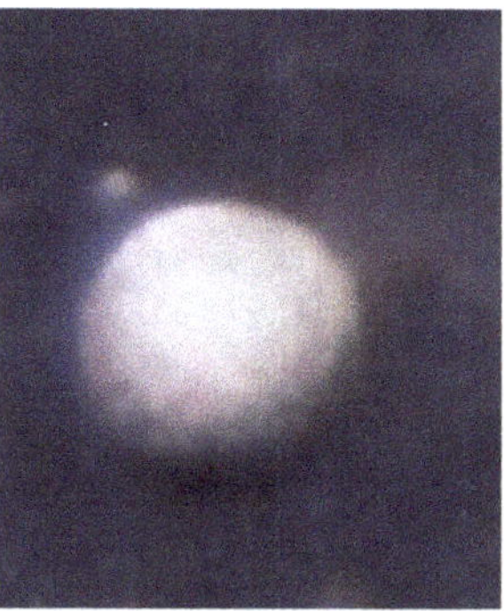

Figure 2.11 Single crystal sphere of InSb

2.5.7.3 DISCUSSION OF THE RESULTS

The structure factors obtained using JANA 2006 [Petříček et al., 2006], as seen in table 2.18, show a close matching of observed and calculated structure factors. The reliability indices are a low indication of the completeness of the refinement procedures. The cell parameters obtained from the JANA 2006 [Petříček et al., 2006] refinement is in close agreement with the reported ones.

Table 2.18 The structure factors of InSb obtained using JANA 2006 software

h	k	l	F_o	F_c	$\sigma(F_o)$
-1	1	1	350.53	347.51	3.57
1	1	1	351.52	348.07	3.58
0	0	2	29.39	17.52	0.97
-2	2	2	19.43	23.95	1.57
0	2	2	463.88	444.13	4.64
2	2	2	19.69	22.86	1.57
-1	1	3	357.88	374.26	3.62
1	1	3	356.39	373.08	3.61
-3	3	3	334.79	347.77	3.58
-1	3	3	347.99	351.05	3.54
1	3	3	351.35	352.49	3.57
3	3	3	336.81	346.38	3.60
0	0	4	452.78	456.80	4.61
-2	2	4	446.97	445.14	4.53
0	2	4	33.19	26.88	0.68
2	2	4	450.99	445.18	4.57
-4	4	4	394.19	391.62	4.53
-2	4	4	33.715	42.43	1.49
0	4	4	432.88	426.85	4.54
2	4	4	33.43	40.89	1.51
4	4	4	392.99	391.83	5.05
-1	1	5	332.34	331.85	3.42
1	1	5	335.29	333.36	3.45
-3	3	5	298.27	278.56	3.23
-1	3	5	317.49	322.54	3.27
1	3	5	315.85	321.05	3.26
-3	3	5	301.02	280.31	3.25
-3	5	5	272.04	302.23	3.36
-1	5	5	283.62	271.89	3.13
1	5	5	285.99	273.63	3.15
3	5	5	271.15	300.76	4.38
-2	2	6	45.20	39.47	1.47
0	2	6	411.82	407.73	4.25
2	2	6	29.02	38.32	1.71
-2	4	6	376.11	374.46	3.89
0	4	6	27.49	35.70	1.72
2	4	6	376.75	374.63	3.88
-1	1	7	287.12	286.49	3.21
1	1	7	282.99	284.86	3.18
-1	3	7	268.56	260.91	2.98
1	3	7	267.42	262.69	3.18

Table 2.19 The structural parameters of InSb obtained using JANA 2006 software

Parameters	With h+k+l=4n+2	Without h+k+l=4n+2
R (obs) (%)	2.92	0.99
*w*R (obs) (%)	4.99	1.49
Extinction (%)	0.01	0.01

REFERENCES

[1] Abrikosov N.H, Shelimova Poluprovodnikovye L.E, materially naosnove soedinenij AIVBVI Nauka (in Russian), Moskva (1975).

[2] Altomare A, Cascarano G, Giacovazzo C, Guagliardi A, J. Appl. Cryst. vol. 27 (1994) pp. 1045–1050.
http://dx.doi.org/10.1107/S002188989400422X

[3] Baldinozzi J, Berar J.F, J. Appl. Crystallogr. vol. 26 (1993) p. 128.
http://dx.doi.org/10.1107/S0021889892009725

[4] Boriseneko V.E (Ed.), Semiconducting Silicides, Springer, Berlin (2000) p.285.
http://dx.doi.org/10.1007/978-3-642-59649-0

[5] Dollase W. A, J. Appl Cryst. vol. 19 (1986) pp. 267–272.
http://dx.doi.org/10.1107/S0021889886089458

[6] Doriane del Frari, Sebastian Deleberto, Nicolas Stein & Jean Marie Lecuire, Thin Solid Films, vol. 483 (2005) p. 240
http://dx.doi.org/10.1016/j.tsf.2004.12.015

[7] Hau K.F, Loo S, Guo F, Chen W, Dyck J.S, Dyck C, Science vol. 303 (2004) p. 818.
http://dx.doi.org/10.1126/science.1092963

[8] Heremans J, Portin D and Thrush C.M, Semicond. Sci. Technol., vol. 40 (1993) pp. 542-549.

[9] Howard C.J, J. Appl. Cryst. vol.15 (1982) pp. 615-620.
http://dx.doi.org/10.1107/S0021889882012783

[10] Holland T.J.B, Redfern S.A.T, Mineralogical Magazine vol. 61 (1997) p. 65.
http://dx.doi.org/10.1180/minmag.1997.061.404.07

[11] Hyun D. P, Prokesa S.M, Twigga M.E, Yong D, and Wang Z. L, Journal of Crystal Growth vol. 304 (2007) pp. 399–401.
http://dx.doi.org/10.1016/j.jcrysgro.2007.03.023

[12] Kajikawa T, Shida Shiraishi K.S, Ohmori T.M, Hirai T, IEEE (1998) p. 362.

[13] Kim Y, DiVenere A, Wong G K L, Kelterson J.B, Cho S, Meyer J.R, J. Appl. Phys. vol. 91 (2002) p. 715.
http://dx.doi.org/10.1063/1.1424056

[14] Kondoh K, Oginuma H and Kimura A, Mater. Trans. vol. 44 (2003) p. 981.
http://dx.doi.org/10.2320/matertrans.44.981

[15] Krivosheeva A.V, Kholod A.N, Shaposhnikov V.L, Semiconductors vol. 36 (2002) p. 496.
http://dx.doi.org/10.1134/1.1478538

[16] LaBotz R.J, Mason D.R, O'Kane D.F, Electrochem. J. Soc. vol. 110 (1963) p. 127.
http://dx.doi.org/10.1149/1.2425689

[17] Lange H, Phys. Stat. Sol. (b) vol. 201 (1997) pp. 3-65.
http://dx.doi.org/10.1002/1521-3951(199705)201:1<3::AID-PSSB3>3.0.CO;2-W

[18] Leong D, Harry M, Reeson K, Homewood K. P, Nature vol. 387 (1997) pp. 686-688.
http://dx.doi.org/10.1038/42667

[19] Madelung O. Semiconductors: Data Handbook, 3rd ed., Springer (2004) pp. 1595-1605.
http://dx.doi.org/10.1007/978-3-642-18865-7

[20] Mahan J.E, Vantomme A, Langouche G, Becker J.P, Phys. Rev. B 54 (1996) p. 16965.
http://dx.doi.org/10.1103/PhysRevB.54.16965

[21] March A, Kristallogr Z. vol. 81 (1932) p. 285.

[22] Miyazaki T, Adachi S, Appl. Phys. vol.70 (1991) pp. 1672-1685.
http://dx.doi.org/10.1063/1.349535

[23] Morris R.G, Redin R.D, Danielson G.C, Phys. Rev. Vol. 109 (1958) pp. 1909.
http://dx.doi.org/10.1103/PhysRev.109.1909

[24] Morten F.D, King R.E, Appl. Optics vol. 4 (6) (1965) pp. 659-668.
http://dx.doi.org/10.1364/AO.4.000659

[25] Noda Y, Kon H, Furukawa Y, Otsuka N, Nisida I.A, Masumoto K, Mater. Trans. JIM vol. 33 (1992) p. 845.
http://dx.doi.org/10.2320/matertrans1989.33.845

[26] Petříček V, Dušek M, Palatinus L, Institute of Physics, Academy of sciences of the Czech republic, Praha (2000).

[27] Petříček V, Dušek M, Palatinus L, Institute of Physics, Academy of sciences of the Czech republic, Praha (2006).

[28] Ravich Yu I, Efimova B.A, Smirnov I.A, Plenum Press, New York, London, (1970).

[29] Redin R.D, Morris R.G, Danielson G C, Phys. Rev. vol. 109 (1958) p. 1916. *http://dx.doi.org/10.1103/PhysRev.109.1916*

[30] Rietveld H. M, J. Appl. Cryst. vol. 2 (1969) pp. 65-71. *http://dx.doi.org/10.1107/S0021889869006558*

[31] Rowe D.M, Thermoelectrics Handbook: Macro to Nano, CRC Press, New York, Section III, Chapter 27 (2006) p. 16.

[32] Satterthwaite C.B, Ure R, Phys. Rev. vol.108 (5) (1957) p. 1164. *http://dx.doi.org/10.1103/PhysRev.108.1164*

[33] Senthil Kumar V, Venkatachalam S, Viswanathan C, Gopal S, Narayandass Sa. K, Mangalaraj D, Wilson K.C, and Vijayakumar K.P, 2005 Cryst. Res. Technol. vol. 40 (6) (2005) pp. 573–578.

[34] Shi W.D, Zhou L, Song S.Y, Yang J.H, Zhang H.J, Adv. Mater. vol. 20 (2008) pp. 1892–1897. *http://dx.doi.org/10.1002/adma.200702003*

[35] Slack G.A, CRC Handbook of Thermoelectrics edited by Rowe D.M.

[36] Stella A, Lynch D.W, J. Phys. Chem. Solids vol. 25 (1964) p. 1253. *http://dx.doi.org/10.1016/0022-3697(64)90023-X*

[37] Sunglae Cho, Antonio DiVenere, George K. Wong, John B. Ketterson, Jerry R. Meyer, J. Appl. Phys. vol. 85 (1999) p. 3655. *http://dx.doi.org/10.1063/1.369729*

[38] Tan J, Proceedings of SPIE 5836 (2005) p. 71. *http://dx.doi.org/10.1117/12.609819*

[39] Thompson P, Cox D. E, Hastings J. B, J. Appl. Cryst. vol. 20 (1987) pp. 79-83. *http://dx.doi.org/10.1107/S0021889887087090*

[40] Touzelhaev M.N, Zhou P, Venkatasubramanian R, Goodson K B, J Appl Phys vol.91 (2001) p. 763. *http://dx.doi.org/10.1063/1.1374458*

[41] Udayashankar N.K, Bhat H.L, 2001, Bull. Mater. Sci. vol. 7 (a) (2001) pp. 744-752.

[42] Uemrau K, Nishida.I.A, Nikkan Kogyo, Thermoelectric semiconductors and their applications (1988) pp. 148-159.

[43] Wertheim G.K, Butler M.A, West K.W, Buchanan D.N.E, Rev. Sci. Instrum. vol. 45 (1974) pp. 1369-1371.
http://dx.doi.org/10.1063/1.1686503

[44] Wyckoff R.W.G, (London) Inter-Science Publishers, vol. I (1963).

[45] Wyckoff R.W.G, (London) Inter-Science Publishers, vol. II (1963).

[46] Wyckoff R.W.G, (London) Inter-Science Publishers, vol. III (1963).

[47] Yamaguchi S, Matsumoto T, Yamazaki J, Kaiwa N, Yamamoto A, Appl. Phys. Lett. vol. 87 (2005) p. 201902.
http://dx.doi.org/10.1063/1.2130390

[48] Young R.A, The Rietveld Method, Oxford University Press (1993).

[49] Zasavitsky E.A, Kantser V.G, Meglei D.F, Semiconductor material conference (2005).

[50] Zou H.L., Rowe D.M., Min G., J. Vac. Sci. Technol. A 19 (2001) pp. 899–903.
http://dx.doi.org/10.1116/1.1354600

CHAPTER III

Results and Discussion on Charge Densities Derived from the Maximum Entropy Method (MEM)

Abstract

The study of the charge density and its distribution inside a unit cell plays a vital role for the better understanding of thermoelectric behavior. The earlier studies on thermoelectric materials revealed that the optimum charge density and the proper distribution of charges at charge centers lead to high figure of merit [Isoda *et al.*, 2006, 2008, Zhang *et al.*, 2008]. Also, the bonding nature of materials characterizes their thermoelectric nature. A high resolution pictorial analysis will give better insight for the study of the electron density distribution and the bonding nature. In this work, high resolution charge density distribution analysis is done using the versatile technique called maximum entropy method (MEM) [Collins, 1982], a tool which can compute three, two and one dimensional electron density distribution inside a unit cell of crystalline materials [Bricogne and Gilmore, 1989, 1990]. The software program VESTA [Momma and Izumi, 2006] provides provision to study the electron density in different planes both qualitatively and quantitatively. Hence, chapter III is devoted to the analysis, results and discussion of charge density distribution using the versatile technique maximum entropy method (MEM) for some selected thermoelectric materials, both in powder form as well as single crystal form.

Keywords

Charge density, MEM, Strategy, Single crystal, Mg_2Si, PbTe, BiSb, Bi_2Te_3, Sb_2Te_3, InSb, SnGeTe

Contents

3.1 INTRODUCTION

The concept of chemical bonding and electron density distribution is by far one of the most useful, and at the same time one of the most difficult to understand processes in all of chemistry [Brown, 2002]. Understanding the internal structure, connectivity in bonds and the pictorial representation of the electron density may give a better idea of the properties of compounds, which lead to the development of new materials, polymers, and other advanced materials. The accurate structure of thermoelectric materials including doped atoms/molecules, nature of doping, effect of doping are always useful for the understanding of the origin of characteristic properties.

Thus, it is important to use a highly efficient technique for the data-analysis of the structure of novel thermoelectric materials. Maximum entropy method (MEM) [Collins, 1982] is one of the best structural analysis methods, based on the information theory. The application of the MEM to crystallography extended the limits of the crystal structure analysis technique [Collins, 1982; Wilkins *et al.*, 1983; Bricogne, 1988; Bricogne and Gilmore, 1990; Sakata and Sato, 1990; Bricogne, 1993; de Vries *et al.*, 1994; Sivia and David, 1994; Papoular and Cox, 1995; Gilmore, 1996; Takata and Sakata, 1996; Marks and Landree, 1998; Burger and Prandl, 1999; Graafsmaa and de Vries, 1999]. The MEM can compute the electron density from a limited number of diffraction data. The electron density, thus computed by MEM, can be visualized as three dimensional and two dimensional high-resolution electron- density distribution images. Even the presence of one electron hydrogen atoms were clearly visualized, if carefully interpreted in the three dimensional as well as two dimensional electron density MEM diagrams [Saravanan, 2009]. This ability of the MEM can be interpreted as "imaging of diffraction data" as shown in the figure 3.1.

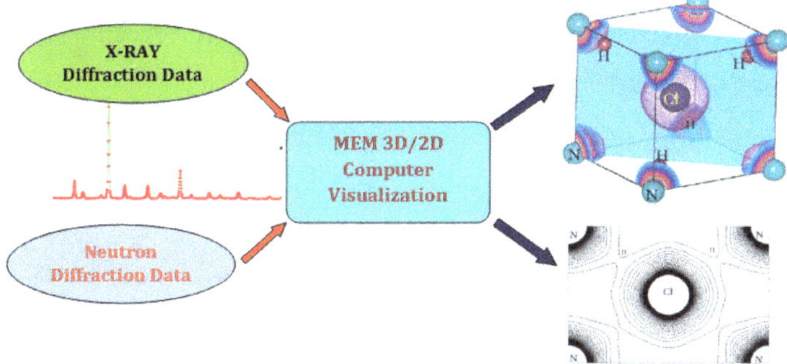

Figure 3.1 Imaging of X-ray and neutron diffraction data using MEM

3.2 ELECTRON DENSITY DISTRIBUTION

The electron density distribution (EDD) is extracted from the set of intensities with the help of precise X-ray diffraction. In most, if not all, centrosymmetric crystals, structure factors can be derived from these intensities without any ambiguity in the phases. The most common method of extracting electron density distribution from an incomplete and noisy set of structure factors is to fit the data to a multipole model [Craven *et al.,* 1987; Hansen and Coppens, 1978; Hirshfeld, 1971, 1977]. The advantages of a multi-pole fit over a direct Fourier synthesis are that a multipole fit allows one to overcome the series termination effect, to filter out the noise from the data that were measured and to extract the static density from the thermally smeared density. A drawback to a multipole fit is that it does not represent the sharp features of the valence charge density due to the extrapolation of available data to the unavailable experimental details of the infinite resolution [Bruning, 1992].

It has been proved in the past few years that the maximum entropy method (MEM) can be used to obtain electron density distribution from an incomplete and noisy set of structure factors [Sakata and Sato, 1990; Sakata *et al.,* 1992; Takata *et al.,* 1993]. The MEM is capable of handling the series termination effect by estimating missing data. The MEM selects the electron density distribution that is closest to a prior electron density distribution or in its absence, closest to a uniform distribution. It is believed that all the features that show up in the electron density distribution are supported by the data and that the MEM gives the least biased results. The major advantages of MEM are that:

- It gives an explicit formulation for the actual electron density rather than normalized density.
- It is the least biased calculation.
- It performs accurately even when available information is limited.
- One can get accurate information about the structure factor.
- Unmeasured and forbidden reflections can be simulated.
- A precise electron density map can be obtained.
- The existence of bonding electrons can be seen.

The charge integration is useful in the quantitative assessment of ionic/covalent/ metallic characters in a solid. When an atom is bonded with its neighbor, the valence charge is expected to be more diffused in comparison with the core charges, which is more concentrated and localized near the nucleus. The integration of charges inside the spheres of radius like covalent/ionic radius will reveal the quantitative measure of the charge transferred from the atom for the bonding process. This can be achieved by carefully selecting spheres with the particular radius and summing over all the charges at the pixels inside the sphere. This charge density is carefully scaled in accordance to get the real charge density. It was found that for a more accurate electron density, the number of pixels should be larger. This will require a high random access memory (RAM) personal computer or supercomputing vector systems. We have analysed the systems by dividing the unit cell into 128x128x128 pixels along the three lattice axes using a personal computer with 4Gb DDR3 RAM.

3.3 STRATEGY FOR MEM CALCULATIONS

The refined structure factors from the Rietveld analysis [Rietveld, 1969] were utilized for the evaluation of MEM charge density using the formalism that Collins (1982) had adopted. The principle and the methodology of MEM calculations have been discussed in the first chapter, which is similar to the one adopted by Sakata and Takata (1996). This procedure deals with a lot of data and hence it has to be run on a supercomputing vector system. However, this report reveals the fact that for higher symmetry systems it is possible to use a high RAM PC with Pentium processor for a MEM calculation, provided each step is done successively and separately. The initial set up for the MEM calculations are listed below:

- Input the parameters h, k, l, F_{real}, F_{ima}, σ (F_{obs}), space group characterized by symmetric cards [International tables for X-ray crystallography, 1974], unit cell parameters a, b, c, and number of reflections.
- The unit cell is to be divided into 128 x 128 x 128 pixels in the case of cubic lattice. In non cubic systems, the unit cell is divided corresponding to the value of the unit cell dimensions a, b and c.
- The asymmetric unit has to be identified using the symmetry of the system.
- Initially, each pixel is to be filled with a charge density of magnitude F_{000}/a_0^3 uniformly, where a_0 is the lattice parameter, F_{000} is the total number of electrons in the unit cell.
- A suitable Lagrangian parameter (λ) has to be chosen.

The convergence of the constraint C to become unity is the final criterion to be noted. When the convergence does not take place, it is necessary to change the value of λ, the Lagrangian parameter and repeat the procedure. It was found that some times, the ill-fated reflections with slightly large $|F_0|-|F_c|$ values along with the incorrect λ do not allow the constraint to converge. Initially, those reflections were set aside and the value of λ was altered, for the convergence to take place quickly. Then those reflections were added again one by one with suitable value of λ, so that convergence took place with a maximum number of reflections and minimum number of iterations.

3.4 MEM REFINEMENTS

During the refinement process, the MEM analysis was performed for all the XRD data sets using Fortran 90 program PRIMA [Ruben and Izumi, 2004], to get a 3D charge density file. The input file contains the cell parameters, space group, pixels, total charge, Lagrange parameter and structural information. The structural information will be the refined results such as miller indices, h k l, the real part of structure factor F_A, the imaginary part of structure factor F_B and the standard deviation between the observed and calculated structure factors, $\sigma(F_{obs})$ values. The phases of the structure factors are calculated using the following relations.

$$F_A = \sum_{j=1}^{n} f_j \mathrm{Cos}\, 2\pi(hx+ky+lz) \tag{3.1}$$

$$F_B = \sum_{j=1}^{n} f_j \sin 2\pi(hx+ky+lz) \tag{3.2}$$

where F_A is the real part of the structure factor and F_B is the imaginary part.

Initially, uniform prior density throughout the unit cell was considered with zeroth order single pixel approximation along with suitable Lagrangian multiplier to get the preference file. The final three dimensional (3D) electron density values were obtained after the refinement process, until the constraint C reaches unity. Finally, with this electron density file the three dimensional (3D) electron densities were plotted using the software package VESTA [Momma and Izumi, 2006]. For the clear understanding of the nature of the bonding in the material, two-dimensional (2D) and one dimensional (1D) distribution of electron density on different lattice planes were plotted. Both qualitative and quantitative electron density studies have been carried out, which has been discussed in this chapter.

The schematic diagram of the MEM analysis is shown in figure 3.2. the final charge density is obtained using MEM refinements with a suitable initial model and the Rietveld [Rietveld, 1969] refined structure details. If the obtained MEM electron density is consistent with the reference model, then one can consider this as the final model. Otherwise, the same process is repeated with a new reference model until the MEM electron density is consistent with the reference model.

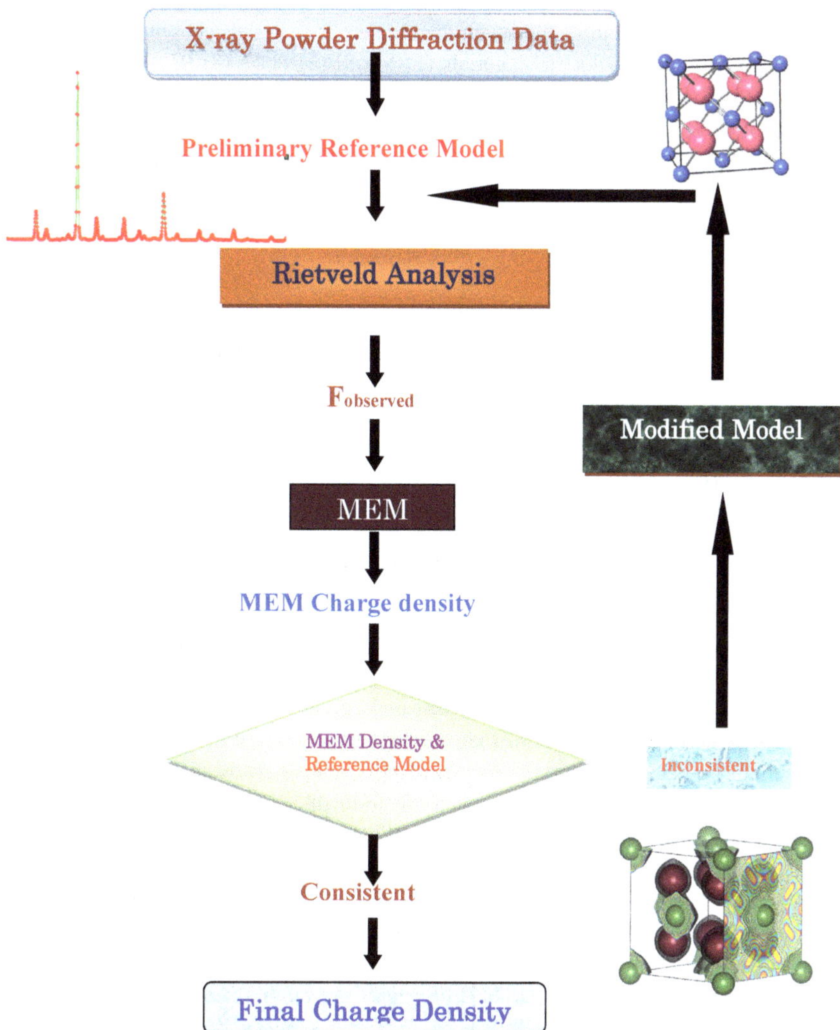

Figure 3.2 Schematic diagram of MEM analysis

3.5 PRESENT WORK IN MEM REFINEMENTS

In the present work, out of the seven thermoelectric materials chosen for this study, the following five thermoelectric materials (1) Mg_2Si (2) PbTe (3) $Bi_{1-x}Sb_x$ with x=0.2 (4) Bi_2Te_3 (5) Sb_2Te_3 have been chosen for the MEM analysis using the powder X-ray diffraction data set. The other two thermoelectric materials chosen (6) $Sn_{1-x}Ge_xTe$ with x=0.12, 0.25 and (7) InSb were analysed using the single crystal data set. The Rietveld [Rietveld, 1969] refined structure factors for various Bragg positions along with the reference model based on crystal symmetry were used as the input for the MEM charge density analysis.

3.5.1 MAGNESIUM SILICIDE (Mg_2Si)

The electronic level charge density analysis and the chemical bonding between the atoms was elucidated for the thermoelectric material Mg_2Si, using the versatile technique MEM. It is well known that the MEM can provide useful information purely from observed structure factor data beyond a presumed crystal structure model used in the pre- Rietveld [Rietveld, 1969] analysis. This visualization ability of a high resolution charge density is useful to construct structure models of thermoelectric materials.

3.5.1.1 MEM REFINEMENTS AND RESULTS

The MEM refinements of Mg_2Si were carried out by dividing the unit cell into 128x128x128 pixels along the three axes a, b and c, the unit cell parameters. The initial electron density at each pixel is fixed uniformly as $F_{000}/a^3 = 0.583$ e/Å^3, where F_{000} is the total number of electrons in the unit cell and "a" is the cell parameter. The Lagrange parameter is suitably chosen so that the convergence criterion C=1 is reached after a minimum number of iterations. The MEM parameters have been given in the table 3.1. The 3D electron density distribution in the form of iso-surface (iso-surface level 0.18 e/Å^3) along with the structure in the unit cell has been represented in figure 3.3. The 3D electron density distribution with the lattice plane (100) in the form of iso-surface (iso-surface level 0.18 e/Å^3) along with the structure in the unit cell has been represented in figure 3.4. The 2D electron density distribution on the (100) and (110) planes has been given in figure 3.5 and figure 3.6 respectively. The one dimensional electron density profiles along [100], [110] and [111] directions are represented in figure 3.7 and figure 3.8 respectively. The numerical values of the electron densities along different directions are given in table 3.2.

Table 3.1 The parameters used and obtained in MEM refinements of Mg_2Si

Parameter	Value
Number of cycles	28
Lagrange parameter (λ)	0.0262
No. of electrons/unit cell (F_{000})	152
R_{MEM} (%)	1.79
wR_{MEM} (%)	1.64

Table 3.2 MEM bond densities of Mg_2Si obtained from one dimensional electron density profiles

Direction	Position (Å) (Distance from origin)	Electron density (e/Å^3)	Comment
[100]	2.288	0.102	Hump
[100]	3.184	0.003	Mid-bond (Si-Si)
[100]	4.079	0.102	Hump
[110]	1.618	0.143	Dip
[110]	2.251	0.190	Hump
[110]	2.880	0.143	Dip
[111]	0.000	98.299	Peak Si
[111]	1.465	0.116	Mid-bond (Si-Mg)
[111]	2.757	59.655	Peak Mg (1)
[111]	4.136	0.108	Dip
[111]	4.566	0.164	Hump
[111]	5.514	0.003	Dip
[111]	6.462	0.164	Hump
[111]	6.893	0.108	Dip
[111]	8.271	59.655	Peak Mg (2)

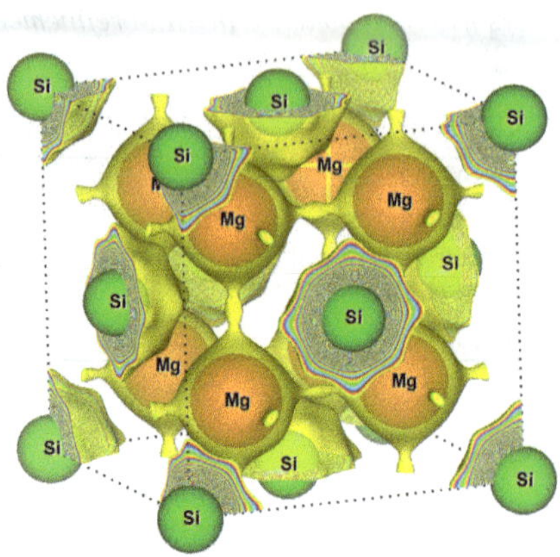

Figure 3.3 3D electron density distribution of Mg_2Si (iso-surface level 0.18 e/Å^3)

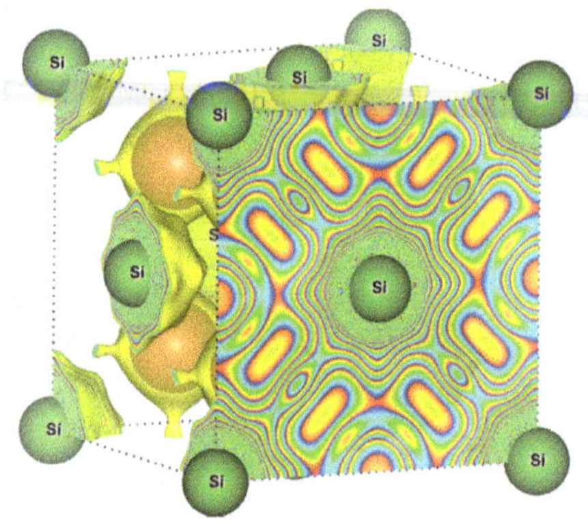

Figure 3.4 3D electron density distribution of Mg_2Si with lattice plane (100)(iso-surface level 0.18 e/Å^3)

Figure 3.5 2D electron density distribution of Mg_2Si on lattice plane (100)[Contour range is from 0.15 e/Å^3 to 2.0 e/Å^3, the step size is 0.03 e/Å^3 and Si atom is at the origin]

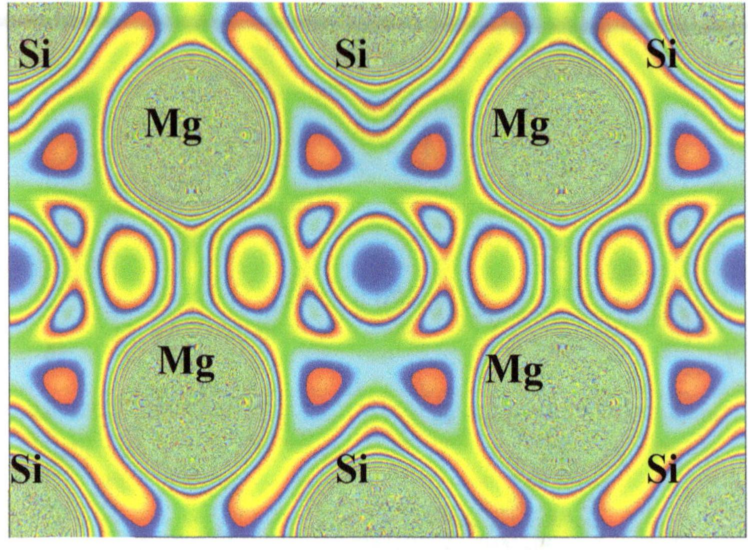

Figure 3.6 2D electron density distribution of Mg_2Si on lattice plane (110)[Contour range is from 0.15 e/Å^3 to 2.0 e/Å^3, the step size is 0.03 e/Å^3 and Si atom is at the origin]

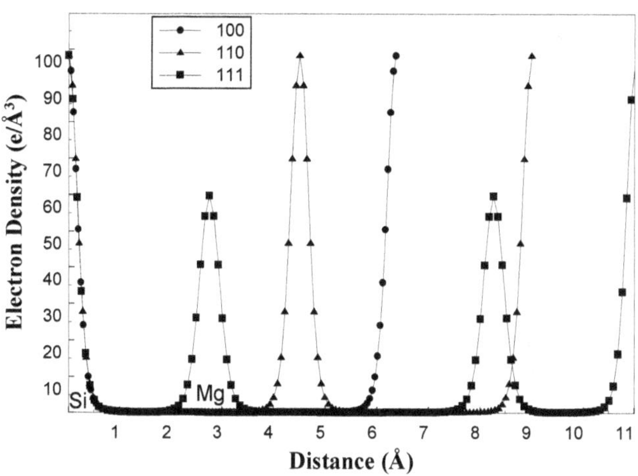

Figure 3.7 One dimensional high MEM electron density profiles of Mg_2Si along [100], [110] and [111] (Si atom is at the origin)

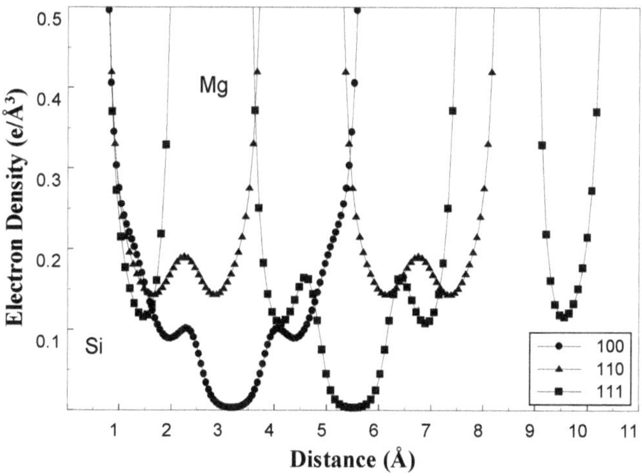

Figure 3.8 One dimensional low MEM electron density profiles of Mg_2Si along [100], [110] and [111] (Si atom is at the origin)

3.5.1.2 DISCUSSION OF THE RESULTS

The 3D MEM electron density distribution shows in both figures 3.3 and 3.4 that the electronic charges in the unit cell are symmetric for both atoms Mg and Si. Figure 3.3 has been constructed superimposing the structural parameters and the radii obtained from the present study and the MEM electron densities. No apparent difference in the size of the core regions is visible due to the comparatively same number of core electrons of Si and Mg. From figure 3.3, it is found that the bonding between Mg atoms (Mg-Mg interaction) is more clearly visible than between Si-Si atoms and Mg-Si atoms. The 2D electron density map on (100) plane, which is shown in figure 3.5 shows the electronic charge distribution of silicon atom, in which the elongations in the valence regions show an attractive character of silicon with its neighbors. Small charge islands are visible at (¼ ¼ ¼) and equivalent positions due to the inner Mg atoms. The two dimensional map on the (110) plane (figure 3.6) clearly shows an attractive character between Si and Mg atoms and covalent character between the Mg atoms. The 1D profiles in figure 3.7 show enhanced peak heights for the Si atom compared to the Mg atom. Figure 3.8 shows nonnuclear maxima, (NNM) along the [110] direction between the Si atoms indicating a covalent character in Si. This covalent bond strength amounts to 0.19 e/Å^3 at a distance of 2.251 Å. The mid-bond electron density between the Si and the Mg atoms along [111] direction is 0.116 e/Å^3 at 1.465 Å. Hence, the bonding between the Si atoms is stronger than that between the Mg atoms and Mg and Si atoms in Mg_2Si.

3.5.2 LEAD TELLURIDE (PbTe)

Lead Telluride (PbTe) is a thermoelectric material with a working temperature of the order of 800 K. The study and pictorial elucidation of electron density and chemical bonding between the atoms in PbTe was carried out using the versatile technique MEM. The real and imaginary parts of the structure factors along with the standard deviation between observed and calculated structure factors obtained using the Rietveld technique [Rietveld, 1969] were used as the inputs for the MEM calculations. The suitable Lagrangian multiplier was selected for the convergence to take place with a minimum number of iterations.

3.5.2.1 MEM REFINEMENTS AND RESULTS

The unit cell of PbTe was divided into 128x128x128 pixels for the MEM refinements. The uniform initial electron density at each pixel was fixed as F_{000}/a^3 =1.999 e/Å^3, where F_{000} is the total number of electrons in the unit cell and "a" is the cell parameter. The MEM parameters along with reliable indices have been given in table 3.3. The 3D electron density distribution in the form of iso-surface (iso surface level 0.6 e/Å^3) in the unit cell has been represented in figure 3.9. The 3D electron density distribution in the form of iso-surface (iso surface level 0.6 e/Å^3) with 2D electron density in the (100) plane has been represented in figure 3.10. The 2D electron density distribution on the (100) and (110) planes has been given in figures 3.11 and 3.12 respectively. The one dimensional electron density profiles along the [100], the [110] and the [111] directions are represented in figure 3.13. The numerical values of the electron densities along different directions are given in table 3.4.

Table 3.3 The MEM parameters along with reliable indices of PbTe

Parameter	Value
Number of cycles	1083
Lagrange parameter (λ)	0.0104
No. of electrons/unit cell (F_{000})	536
R_{MEM} (%)	2.00
wR_{MEM} (%)	2.13

Table 3.4MEM bond densities of PbTe obtained from one dimensional electron density profiles

Direction	Position (Å)	Electron density (e/Å^3)
[100]	0.000 (Te atom)	688.35
[100]	1.618 (Mid-bond)	0.444
[100]	3.229 (Pb atom)	795.96
[110]	2.282	0.013
[111]	2.796	0.003

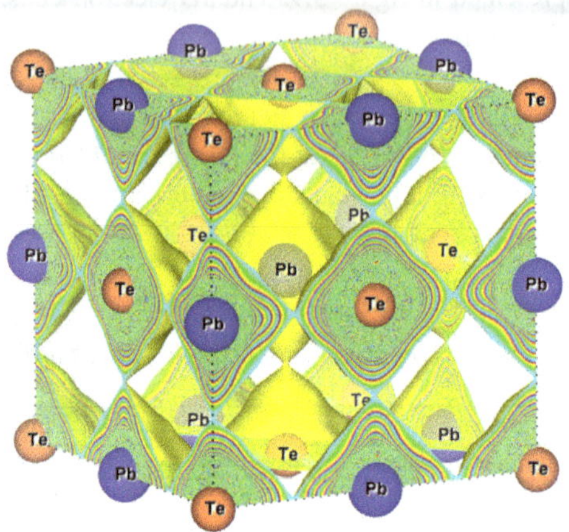

Figure 3.9 3D MEM electron density of PbTe in the unit cell (iso-surface level 0.6 e/Å^3, Te atom is at the origin)

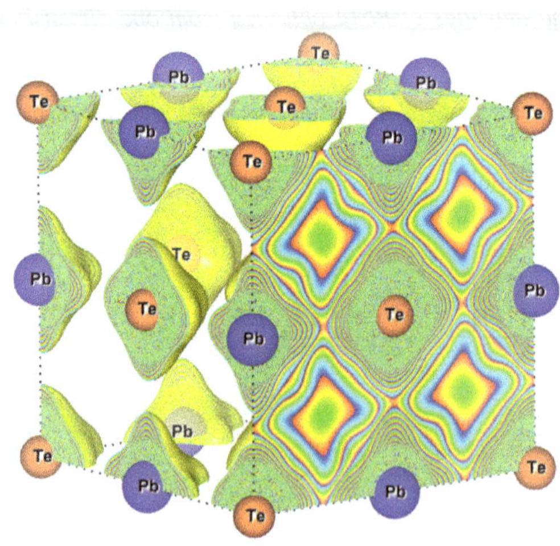

Figure 3.10 3D MEM electron density of PbTe in the unit cell along with 2D electron density in (100) plane (iso-surface level 0.6 e/Å3, Te atom is at the origin)

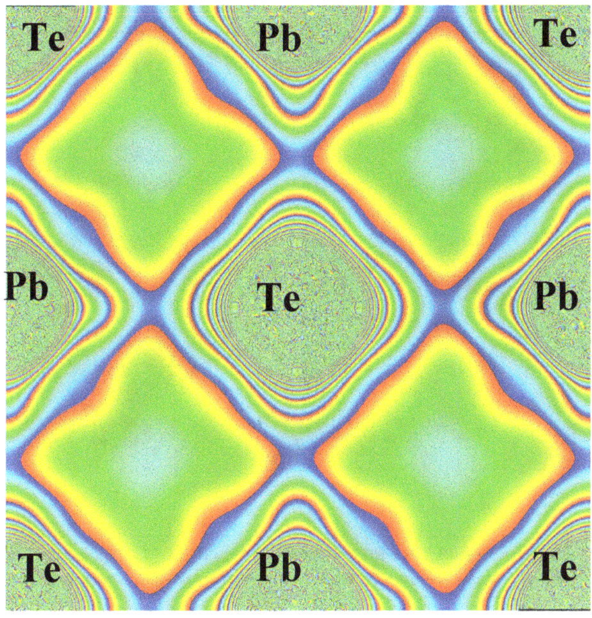

Figure 3.11 2D MEM electron density map of PbTe on (1 0 0) plane (Contour range is 0.045 e/Å3-16.0 e/Å3, the step size is 0.4 e/Å3, Te atom is at the origin)

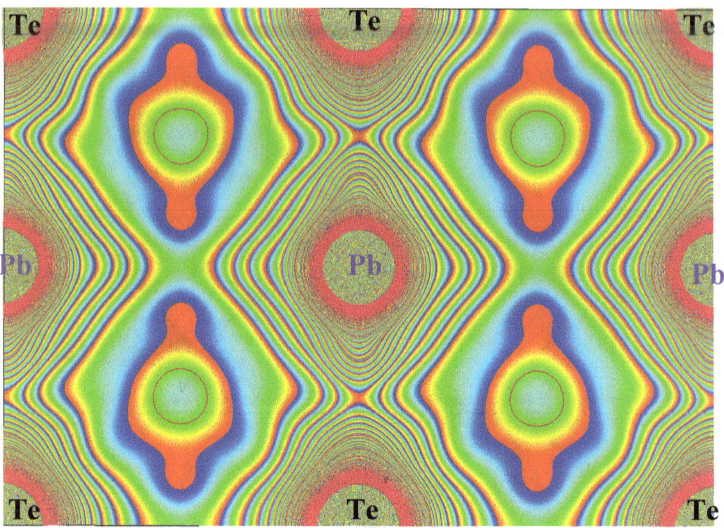

Figure 3.12 2D MEM electron density map of PbTe on (1 1 0) plane (Contour range is 0.045 e/Å3 - 0.4 e/Å^3, the step size is 0.4 e/Å^3, Te atom is at the origin)

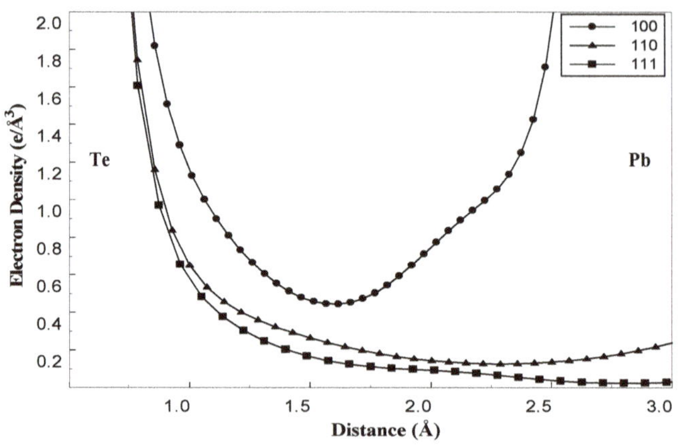

Figure 3.13 One dimensional MEM electron density profiles of PbTe along [100], [110] and [111] (Te atom is at the origin)

3.5.2.2 DISCUSSION OF THE RESULTS

The 3D MEM electron density distribution shows (figure 3.9) that the electronic charges in the unit cell are symmetric for both atoms Pb and Te. The Pb atom has more pronounced elongation of the valence charges towards the Te atom along all the nearest neighbor directions. The 2D electron density map on the (100) plane (figure 3.11) shows more dense electron densities of Pb than Te. The (110) map (figure 3.12) clearly shows a more ionic character indicated by voids all over the plane except atomic positions. The 1D profile in figure 3.14 shows no NNM (nonnuclear maxima) along all three directions. But, the electron density along the mid-bond position is not too low (0.444 e/Å^3) indicating covalent interactions between the Pb and the Te atoms. Hence, the bonding in PbTe is supported to be more ionic than covalent. The mid-bond electron density of 0.444 e/Å^3 occurs at a position of 1.618 Å along the [100] direction. The minimum density along the [111] direction is 0.003 e/Å^3, whereas the [110] direction gives a minimum density of 0.013 e/Å^3 at a distance of 2.282 Å. This position is exactly at 1/4th the face diagonal distance of the (100) plane. *i.e,* the mid bond position along the [110] direction connecting the Te and the Pb atoms. The slightly increased density value may be due to the interaction of the Te and the Pb atoms along the [110] direction. This interaction is visible in the PDF analysis as well.

3.5.3 BISMUTH DOPED WITH ANTIMONY (Bi_{80} Sb_{20})

The refined structure factors from the Rietveld [Rietveld, 1969] formalism were utilized for the MEM calculations for both the parental elements Bi, Sb and the compound $Bi_{80}Sb_{20}$. The preliminary reference model corresponding to Bismuth was considered for $Bi_{80}Sb_{20}$, considering perfect doping of the antimony atoms (Sb) on the bismuth atomic sites (Bi).

3.5.3.1 MEM REFINEMENTS AND RESULTS

The refinements were carried out by dividing the unit cell into 128×128×128 pixels. The MEM refinement parameters along with reliability index factor have been given in table 3.5. The 3D electron density of Bi, Sb and $Bi_{80}Sb_{20}$ are shown in figure 3.14. Figure 3.15 shows the (110), the (024) and the (001) planes in the unit cell of Bi along with the electron densities. The 2D electron density distributions of Bi, Sb and $Bi_{80}Sb_{20}$ on the (110) and the (024) plane have been plotted for comparison under the same plotting parameters in figure 3.16 and figure 3.17 respectively. The one dimensional electron density profiles along the [001] and the [024] directions are represented in figure 3.18 and figure 3.19 respectively. The numerical values of the electron densities along different directions are given in table 3.6.

Table 3.5 MEM Parameters of Bi, Sb and $Bi_{80}Sb_{20}$

Parameter	Bi	Sb	$Bi_{80}Sb_{20}$
Number of cycles	133	132	4451
Lagrange parameter (λ)	0.116	0.1155	0.0015
F_{000}	498	306	459
R_{MEM} (%)	2.3313	2.4594	0.5324
wR_{MEM} (%)	2.888	2.7253	0.2976

Table 3.6 MEM electron densities along different directions of the unit cell of Bi, Sb and $Bi_{80}Sb_{20}$ [ED stands for electron density]

	Bi		Sb		$Bi_{80}Sb_{20}$	
Direction	Distance (Å)	ED (e/Å^3)	Distance (Å)	ED (e/Å^3)	Distance (Å)	ED (e/Å^3)
[110]	1.246	0.157	1.013	0.106	0.740	0.181
[110]	1.459	0.154	1.452	0.089	1.445	0.104
[110]	2.279	0.273	2.161	0.129	2.256	0.247
[110]	3.097	0.153	2.870	0.089	3.067	0.104
[024]	1.514	0.094	1.439	0.035	1.502	0.281
[024]	12.108	0.273	11.515	0.129	10.888	0.292
[024]					13.139	0.292

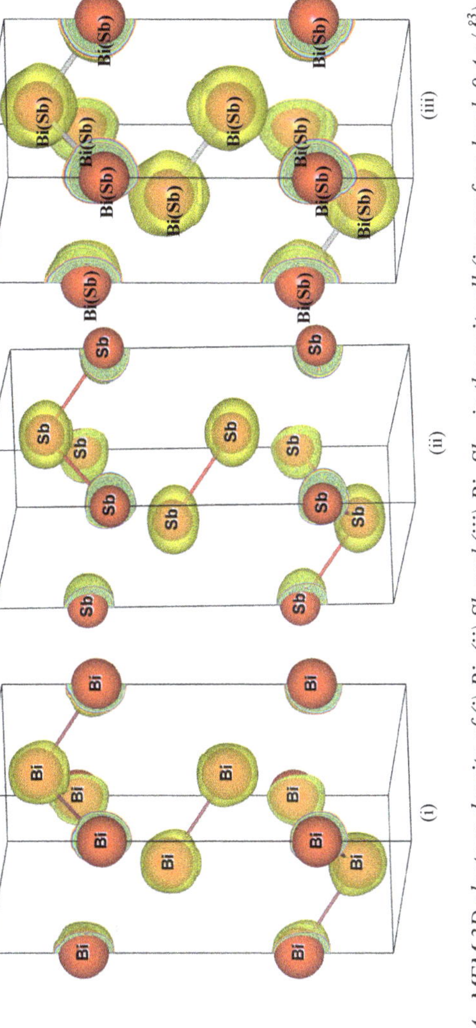

Figure 3.14 MEM 3D electron density of (i) Bi, (ii) Sb and (iii) $Bi_{80}Sb_{20}$ in the unit cell (iso surface level 0.4 e/Å^3)

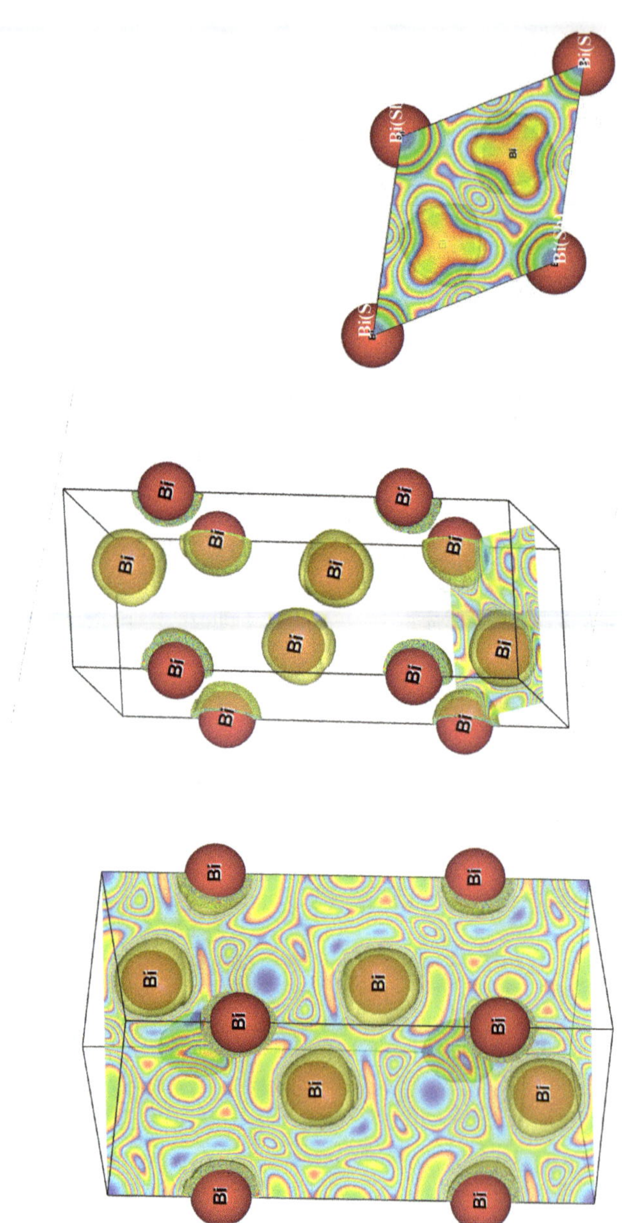

Figure 3.15 MEM 2D electron density on (110), (024) and (001) planes in the 3D unit cell of Bi

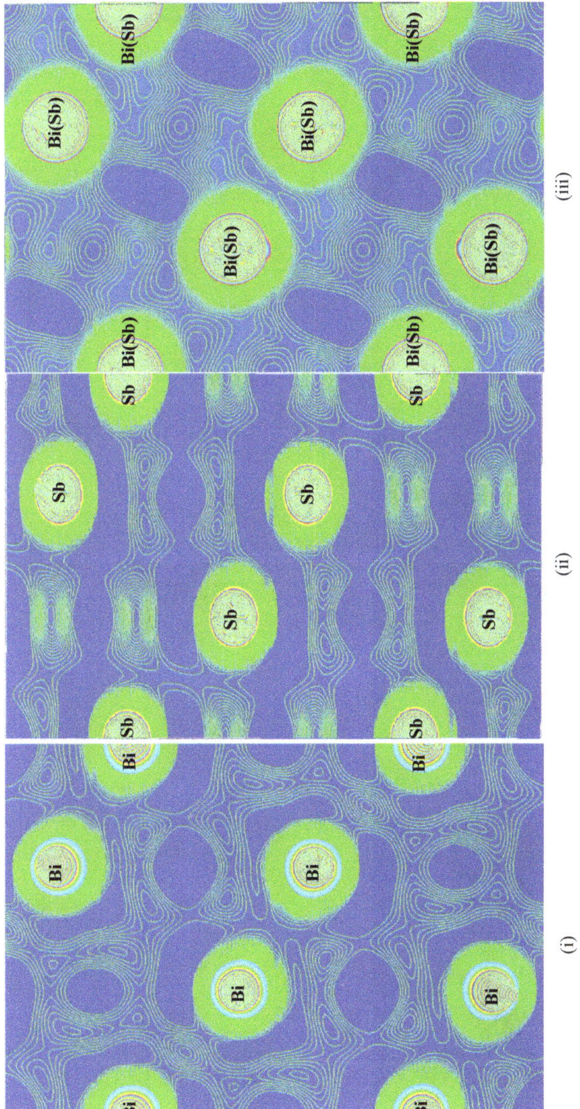

Figure 3.16 MEM 2D electron density map of (i) Bi, (ii) Sb and (iii) $Bi_{80}Sb_{20}$ on the (110) plane (Contour range is from 0.1 e/Å^3 to 4.0 e/Å^3, The interval is 0.02 e/Å^3)

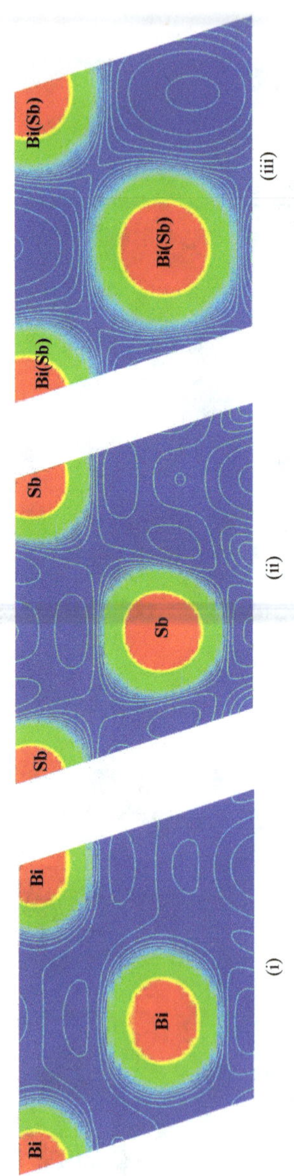

Figure 3.17 MEM 2D electron density map of (i) Bi (ii) Sb and (iii) Bi80Sb20 on (024) plane (Contour range is from 0.0 e/Å3 to 5.0 e/Å3, the interval is 0.04 e/Å3)

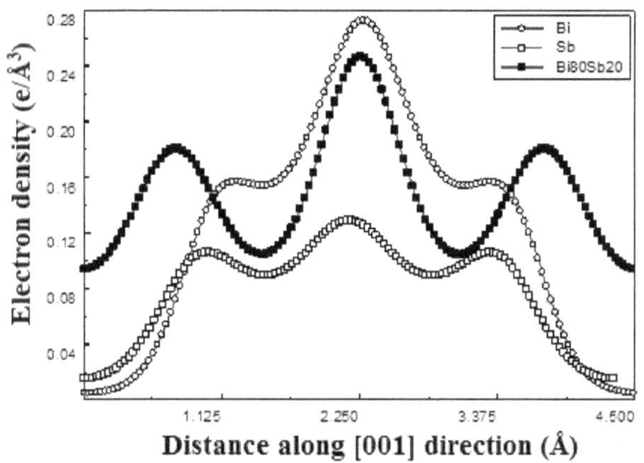

Figure 3.18 One dimensional MEM electron density profiles of Bi, Sb and $Bi_{80}Sb_{20}$ along [001] direction

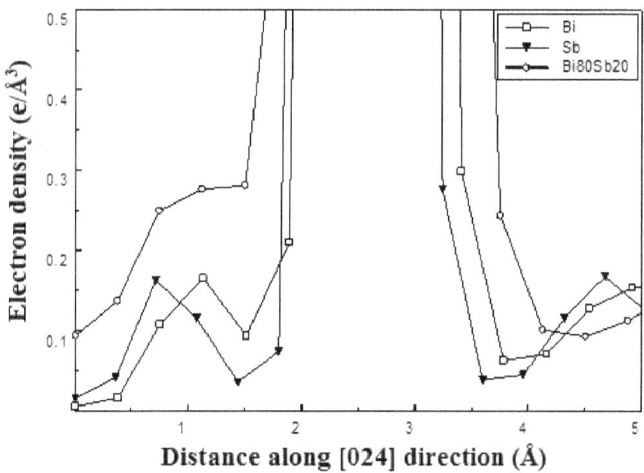

Figure 3.19 One dimensional MEM electron density profiles of Bi, Sb and $Bi_{80}Sb_{20}$ along [024] direction

3.5.3.2 DISCUSSION OF THE RESULTS

The 3D electron density distributions of Bi, Sb and $Bi_{80}Sb_{20}$ in the unit cell in figure 3.14 explicitly show the enhancement of electron density in $Bi_{80}Sb_{20}$ due to the single phase formation of Bi and Sb. This enhancement of the electron density in $Bi_{80}Sb_{20}$ results in a high electrical conductivity for this thermoelectric material. A phonon glass electron crystal (PGEC) material features cages (or tunnels) in the crystal structure inside which reside atoms small enough to rattle, i.e., to create dynamic disorder. This situation produces a phonon damping effect which results in a reduction of the solid lattice thermal conductivity. From figure 3.16, the 2D MEM electron density maps of Bi, Sb on the (110) plane show that though there are open channels (voids) in Sb and Bi for higher lattice vibrations, there is much less space for the atoms to contribute lattice vibration in $Bi_{80}Sb_{20}$. This is one way to increase the ZT by minimizing lattice thermal conductivity, while retaining good electrical and thermopower properties. It is also evident from table 3.5 that the static Debye–Waller factor B for $Bi_{80}Sb_{20}$ is nearly three times higher than the host atom Bi due to the substitution of the impurity Sb. This may result in the increase in atomic size in $Bi_{80}Sb_{20}$, which is explicitly seen in figure 3.16 and figure 3.17, which in turn, results in lattice vibration.

3.5.4 BISMUTH TELLURIDE (Bi_2Te_3)

The MEM refinements were carried out by dividing the unit cell into 64x64x448 pixels. The initial electron density at each pixel is fixed uniformly as F_{000}/a_0^3 = 0.0037 e/Å^3 and 1.106 e/Å^3, where F_{000} is the total number of electrons in the unit cell and a_0 is the cell parameter. The Lagrange parameter λ=0.6358 is chosen for the convergence with a minimum number of iterations.

3.5.4.1 MEM REFINEMENTS AND RESULTS

The MEM parameters have been given in table 3.7. All the electron density diagrams are visualized within the unit cell boundary x=y=0 to 1 and z=0.5 to 1. The 3D electron density distribution in the form of iso-surface (iso-surface level 2.0 e/Å^3) in the unit cell is represented in figure 3.20. The 2D electron density on the (100) and the (110) plane superimposed on a 3D structure with the unit cell is visualized in figure 3.21. The 2D electron density distribution on the (100) and the (110) planes in the contour range (0.2-1.0 e/Å^3) with interval 0.08 e/Å^3 are given in figures 3.22 and 3.23 respectively. The one dimensional electron density profiles along the [100], the [110] and the [111] directions are represented in figures 3.24 and 3.25. The numerical values of the electron densities along different directions are given in table 3.8.

Table 3.7 MEM Parameters of Bi_2Te_3

Parameter	Value
Number of cycles	2180
Prior density (e/ Å^3)	1.8916
Lagrange parameter (λ)	0.06358
R_{mem}(%)	0.05009
wR_{mem}(%)	0.05070
Resolution (Å/pixel)	1835008(64x64x448)

Table 3.8 MEM electron densities along different directions of the unit cell of Bi_2Te_3

Direction	Atoms	Position (Å)	Electron Density (e/Å^3)
[001]	Te	6.36	392.17
[001]	Bi	12.10	2410.83
[010]	Te	0.06	273.22
[010]	Bi	2.25	0.84

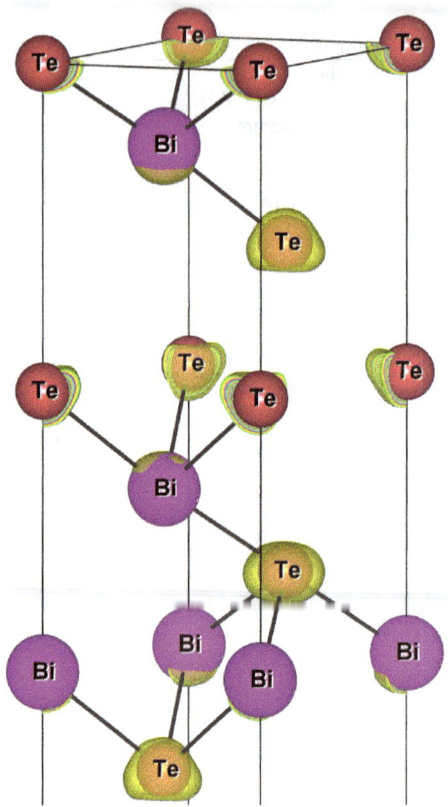

Figure 3.20 3D electron density distribution in the form of iso-surface (iso-surface level 2.0 e/Å^3)

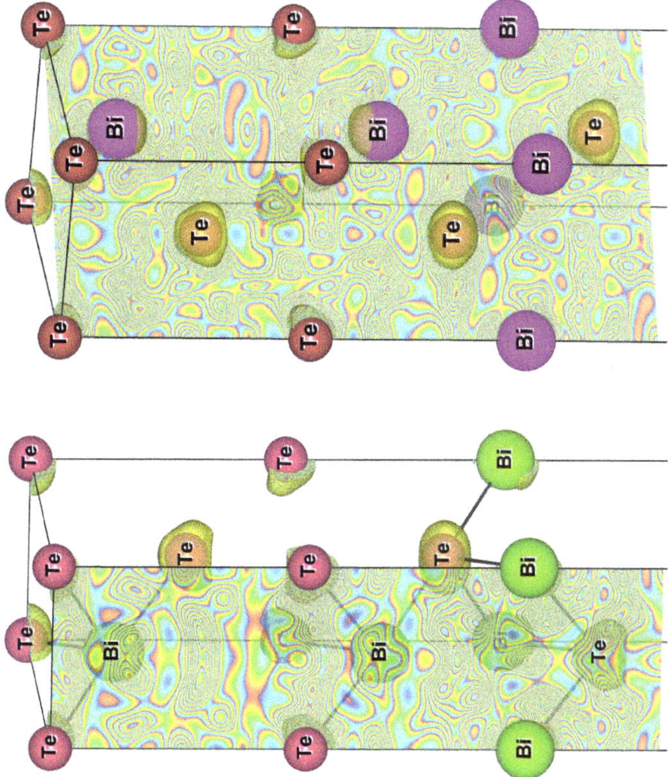

Figure 3.21 2D electron density on (100) and (110) plane superimposed on 3D structure

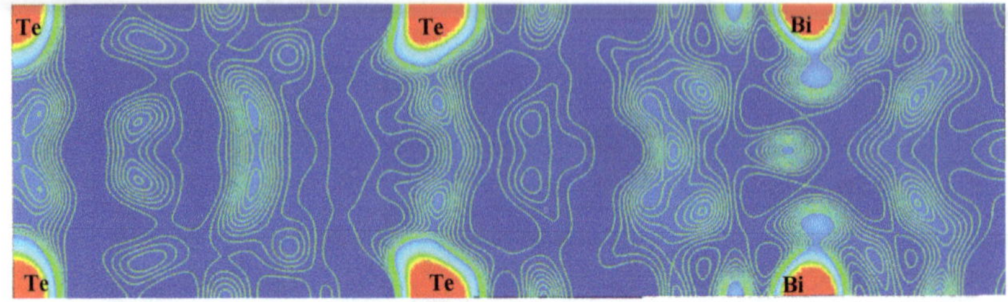

Figure 3.22 2D electron density distribution on the (100) plane [Contour range (0.2 – 1.0 e/Å^3) with interval 0.08 e/Å^3]

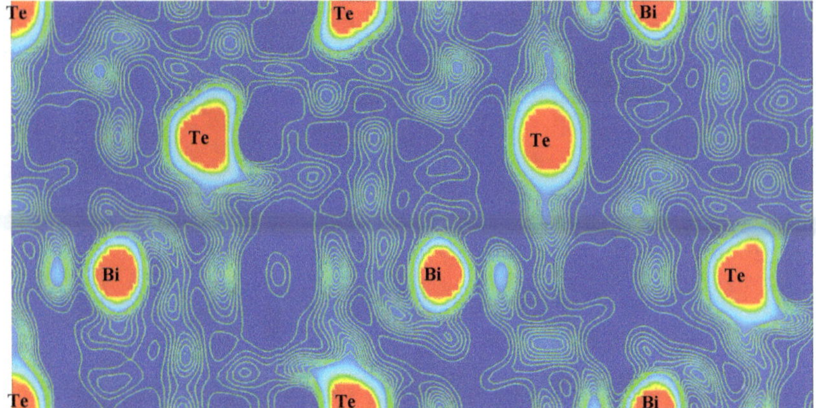

Figure 3.23 2D electron density distribution on the (110) plane [Contour range (0.2 – 1.0 e/Å^3) with interval 0.08 e/Å^3]

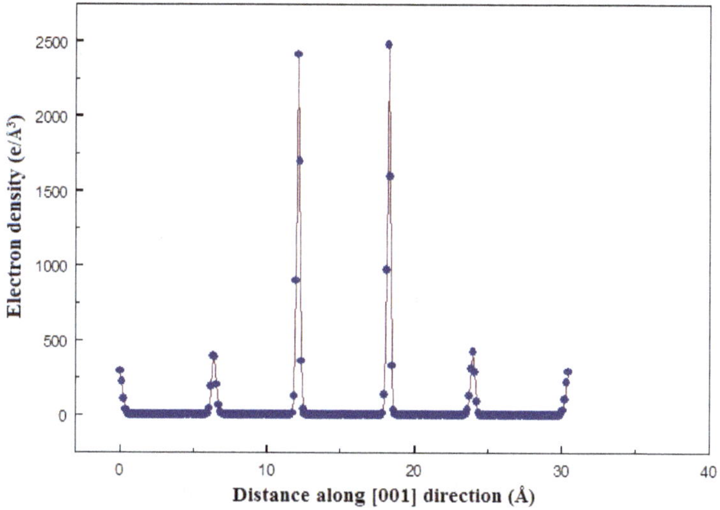

Figure 3.24 One-dimensional profile of high charge density distribution along [001] direction

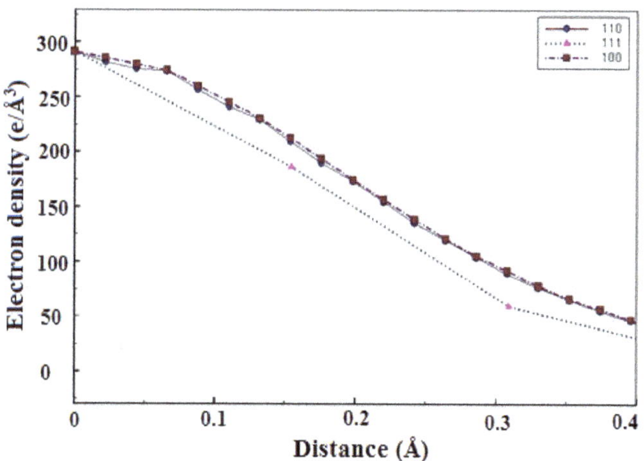

Figure 3.25 One-dimensional MEM high electron density distribution along the [110], the [111] and the [100] directions

3.5.4.2 DISCUSSION OF THE RESULTS

The total electron density distribution in the system is visualized in figure 3.20. The electron density distribution around Bi as well as Te atoms in the form of contour lines shows covalent character. From figure 3.21, it is found that the interactions between the Te atoms are predominant than between the Bi atoms and the interactions between the Bi and the Te atoms. The same nature is also visible in the 2D representation (figure 3.22) on the (100) plane. Also, the 2D low electron density region on the (110) plane is shown in figure 3.23. Figure 3.21 represents three dimensional views, sliced in the (100) and the (110) planes and two dimensional views in the same plane respectively. One can see the variations of electron densities between bismuth and antimony atoms in the (110) plane (figure 3.23). Figure 3.24 shows the one dimensional electron density distribution along the [001] direction, which reveals a variation in density between the bismuth and the tellurium atoms. From table 3.21, one can see the variation in electron density of the bismuth atom which reveals the fact that in the [001] direction, it doesn't share its electron density with its neighbors, hence gaining a maximum value. In the [110] direction (figure 3.25), the bismuth atoms share the electron density with the neighbors and hence gaining a lower density value.

3.5.5 ANTIMONY TELLURIDE (Sb_2Te_3)

The MEM electron densities were computed using the unit cell of Sb_2Te_3 divided into 64×64×448 pixels along the three lattice parameters a, b and c, by considering the rhombohedra structure. The initial electron density at each pixel is set to be F_{000}/a^3 = 1.611 e/A^3. The MEM parameters along with reliable indices are given in table 3.9. The electron density distribution and the bonding between antimony and telluride atoms in antimony telluride are determined using two dimensional maps and one-dimensional profiles constructed using the calculated MEM electron densities.

3.5.5.1 MEM REFINEMENTS AND RESULTS

The numerical values of the maximum electron densities along the [001] and the [010] directions are given in table 3.10. Figure 3.26 represents the three dimensional view of the tellurium and the antimony atoms inside the unit cell. The 2D view of Sb_2Te_3 on the (010) plane is superimposed on the 3D structure, which is represented in figure 3.27. The 2D visualization of Sb_2Te_3 along the plane (010) is shown in figure 3.28 with contour range (0.0 - 1.0 e/Å^3) and with contour interval 0.05 e/Å^3. The 3D view with iso surface level 3.0 e/Å^3 and 2D view with contour range 0.0 - 1.0 e/Å^3 with interval 0.05 e/Å^3, on the plane (001) are shown in figure 3.29 and 3.30 respectively. One dimensional MEM electron density along different direction of the unit cell is given in figure 3.31.

3.5.5.2 DISCUSSION OF THE RESULTS

The 3D and 2D representations of the electron densities explicitly show the layered structure, which supports the enhanced thermoelectric behavior in this material with a moderate figure of merit. The 3D electron density distribution in the system is visualized in figure 3.26. The electron density distribution around the Sb as well as the Te atoms in the form of contour lines shows covalent character. From figure 3.27, it is found that the interactions between the Te atoms is found to be predominant than between the Sb atoms and the interactions between the Sb and the Te atoms. The same nature is also visible (figure 3.30) in the 2D representation in (001) plane.

Table 3.9 MEM parameters along with reliable indices of Sb_2Te_3

Parameter	Value
Number of cycles	2540
Prior density (e/Å^3)	1.61097
Lagrange parameter (λ)	0.07078
R_{MEM} (%)	0.12101
wR_{MEM} (%)	0.10932
Resolution (Å/pixel)	1835008(64x64x448)

Table 3.10 MEM electron densities along different directions of the unit cell of Sb_2Te_3

Direction	Atoms	Position (Å)	Electron Density (e/Å^3)
[001]	Te	6.43	532.64
[001]	Sb	12.09	537.42
[010]	Te	0.00	559.51

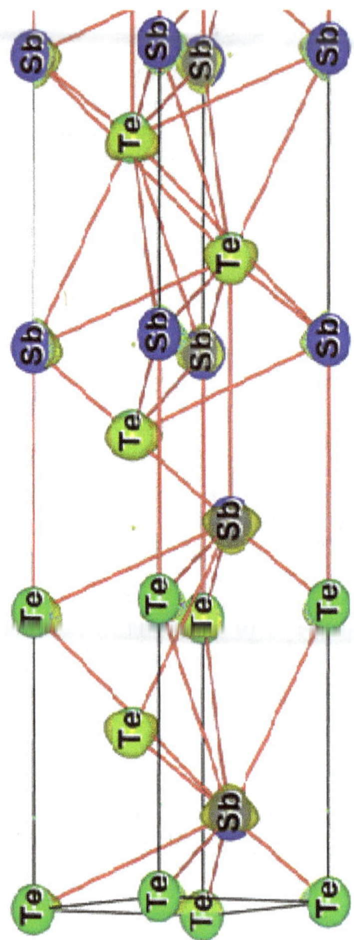

Figure 3.26   3D electron density distribution in the form of iso-surface (iso-surface level 3.0 e/Å³)

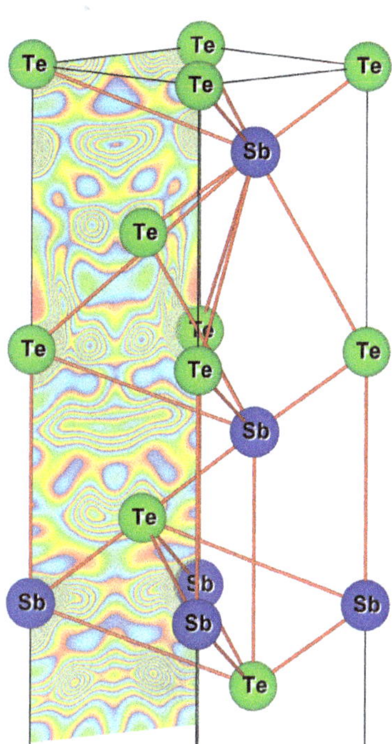

Figure 3.27 3D electron density distribution in the form of iso-surface (iso-surface level 3.0 e/Å^3)

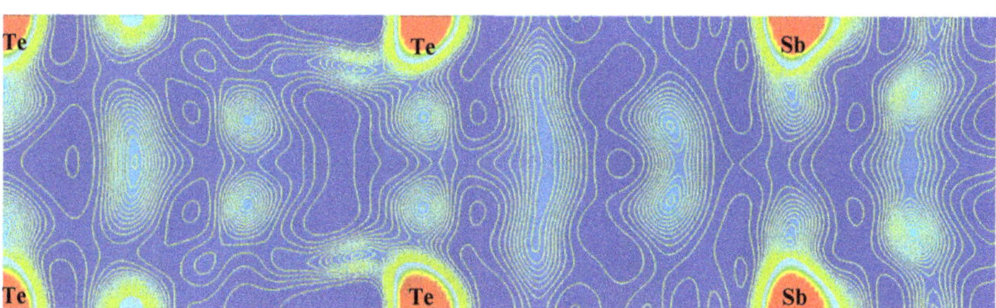

Figure 3.28 2D electron density distribution on the (010) plane in the contour range (0.0 - 1.0 e/Å^3) with interval 0.05 e/Å^3

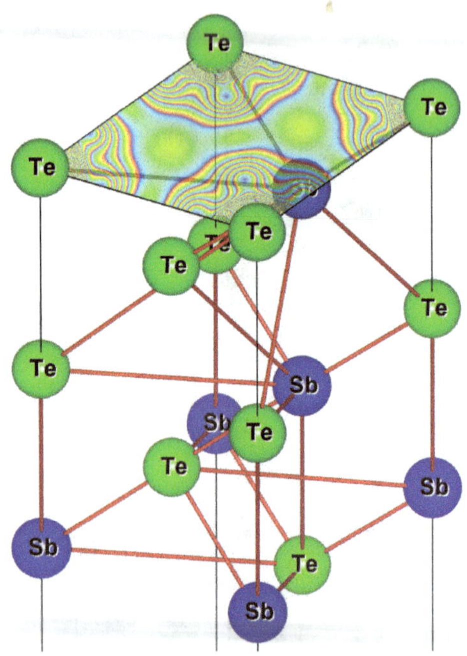

Figure 3.29 2D electron density on (001) superimposed on 3D structure

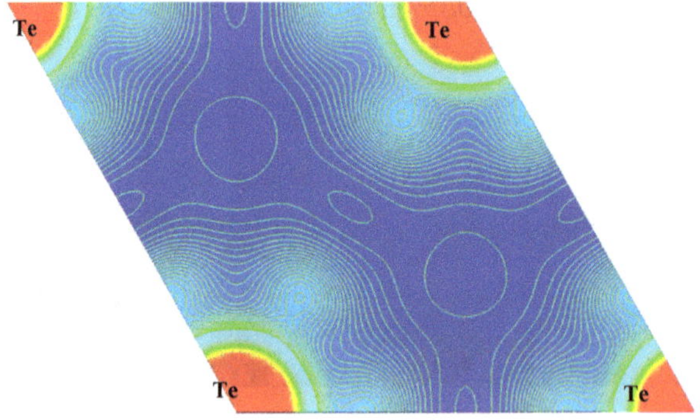

Figure 3.30 2D electron density distribution on the (001) plane in the contour range (0.0 - 1.0 e/Å^3) with interval 0.05 e/Å^3

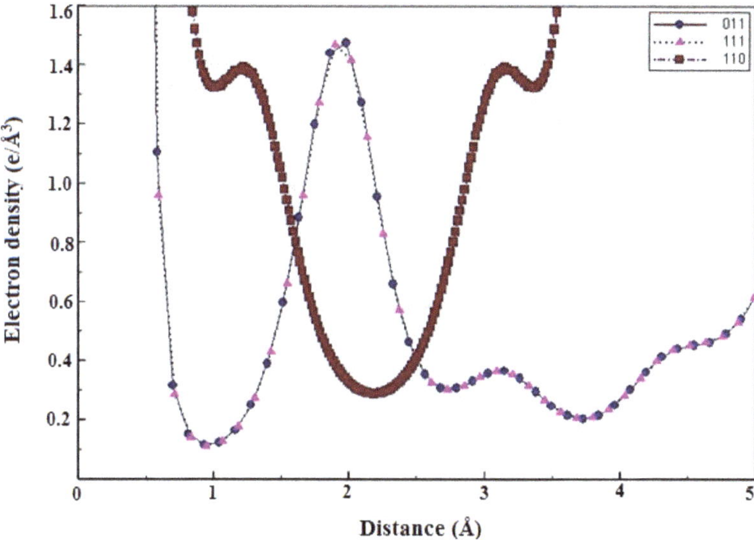

Figure 3.31 One-dimensional MEM high electron density distribution along [011], [111] and [110] directions

3.5.6 $Sn_{1-x}Ge_xTe$ SINGLE CRYSTALS

The electron density analysis and the nature of chemical bonding between the atoms were elucidated for the thermoelectric material SnTe doped with germanium atoms (Ge). Though this material has been used for a long time in space applications, the electron density distribution analysis was nowhere found in the current literature. The MEM computations were carried out using the Rietveld [Rietveld, 1969] refined structure factors. The unit cell was divided into 128x128x128 pixels along the three axis a, b and c, the unit cell parameters. The initial electron density at each pixel was fixed uniformly as F_{000}/a^3 = 1.6285 e/Å^3 for $Sn_{0.75}Ge_{0.25}Te$ and 1.6187 e/Å^3 for $Sn_{0.88}Ge_{0.12}Te$ respectively, where F_{000} is the total number of electrons in the unit cell and "a" is the cell parameter. The Lagrange parameter was suitably chosen so that the convergence criterion C=1 was reached after a minimum number of iterations.

3.5.6.1 MEM REFINEMENTS AND RESULTS

The MEM parameters are given in table 3.11 and the numerical values of the electron densities along different directions are given in table 3.12. The 3D electron density distribution in the form of iso-surface (iso-surface level 0.45 e/Å^3) along with the structure in the unit cell is represented in figure 3.32 for both systems. The 2D electron density distribution for both systems on the lattice planes (100) and (110), in the form of iso-surface (iso-surface level 0.33 e/Å^3) along with the 3D structure in the unit cell is represented in figures 3.33 and 3.34 respectively. The 2D electron density distributions on the (100) and the (110) planes in the contour range (0-15 e/Å^3) with the interval 0.2 e/Å^3 are given in figures 3.35 and 3.36 respectively. The one dimensional electron density profiles along the [100], the [110] and the [111] directions are represented in figure 3.37.

Table 3.11 MEM Parameters of Sn1-xGe.xTe with x=0.12, 0.25

Parameter	$Sn_{0.88}Ge_{0.12}Te$	$Sn_{0.75}Ge_{0.25}Te$
Number of cycles	1301	1154
Lagrange parameter (λ)	0.023	0.022
F_{000}	399	390
R_{MEM} (%)	3.04	2.22
wR_{MEM} (%)	3.46	2.88

Table 3.12 MEM electron densities along different [h k l] directions of the unit cell of Sn1-xGe.xTe with x=0.12, 0.25 (ED stands for electron density)

Direction	$Sn_{0.88}Ge_{0.12}Te$		$Sn_{0.75}Ge_{0.25}Te$	
	Distance (Å)	ED (e/Å^3)	Distance (Å)	ED (e/Å^3)
[100]	0.000	578.85	0.000 (Sn atom)	258.18
[100]	3.135	252.34	3.105 (Te atom)	460.88
[100]	1.469	0.3459	1.553 (Mid-bond)	0.36
[110]	1.385	0.3021	1.372	0.28
[111]	2.715	0.1407	2.017	0.17

F000 = Number of electrons in the unit cell.

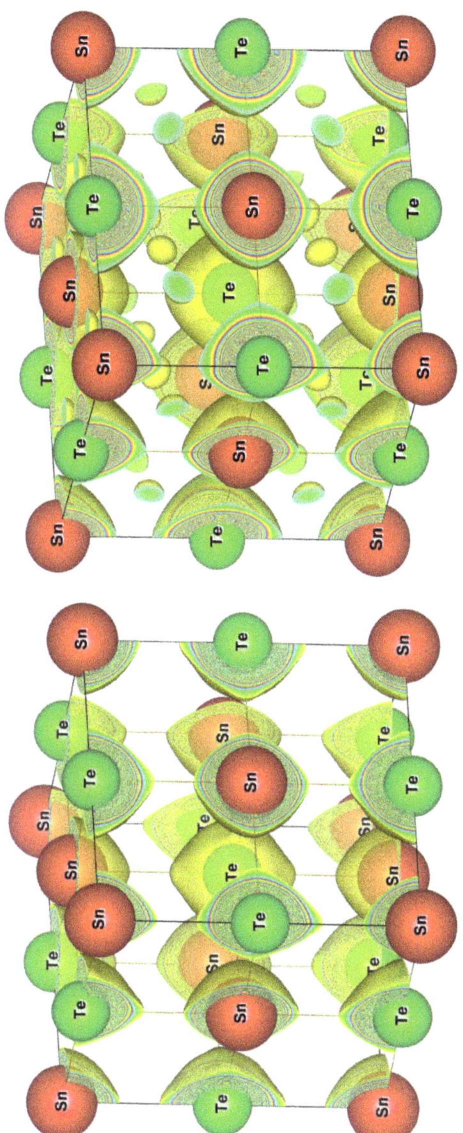

Figure 3.32 MEM 3D iso-surface of the electron density of $Sn_{0.75}Ge_{0.25}Te$ and $Sn_{0.88}Ge_{0.12}Te$ in the structural unit cell (iso-surface level 0.45 e/$Å^3$)

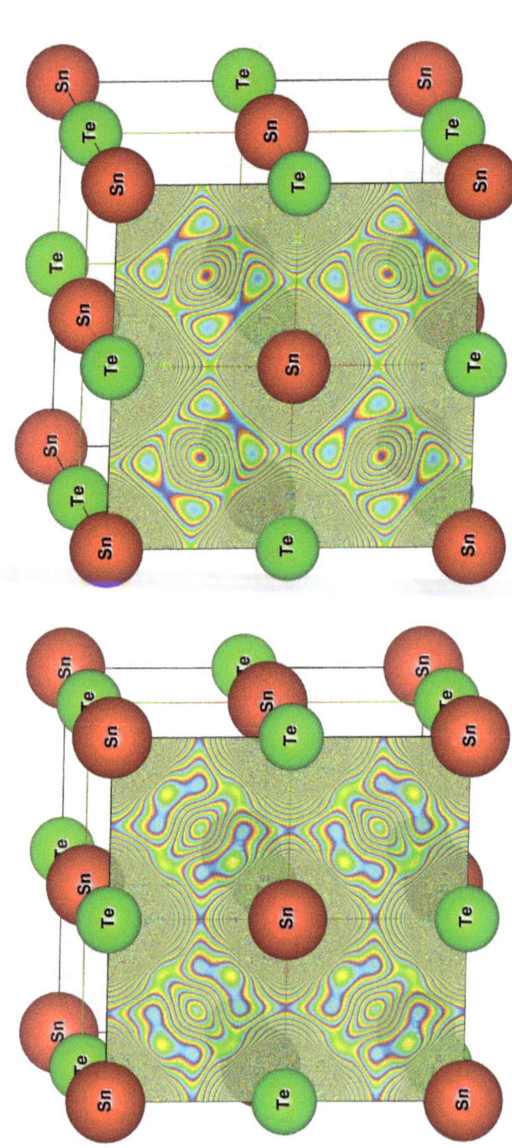

Figure 3.33 MEM 2D electron density on the (100) plane of $Sn_{0.75}Ge_{0.25}Te$ and $Sn_{0.88}Ge_{0.12}Te$ in the structural unit cell (iso-surface level 0.33 e/Å^3)

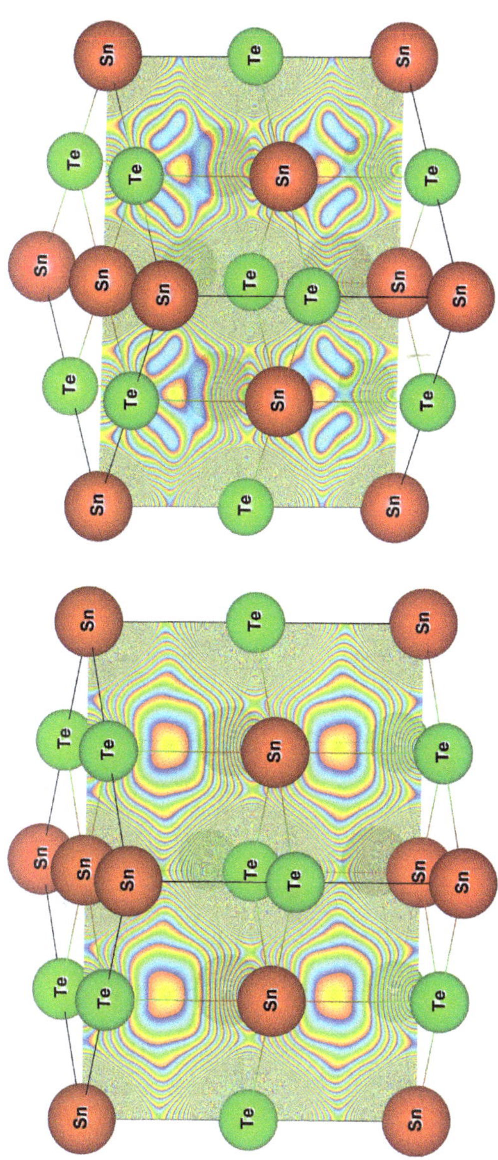

Figure 3.34 MEM 2D electron density on the (110) plane of $Sn_{0.75}Ge_{0.25}Te$ and $Sn_{0.88}Ge_{0.12}Te$ in the structural unit cell (iso-surface level 0.33 e/Å^3)

Figure 3.35 MEM 2D electron density map of $Sn_{0.75}Ge_{0.25}Te$ *and* $Sn_{0\text{-}88}Ge_{0.12}Te$ *on (100) plane, Contour range is (0-15) e/Å*^{3} *with step size is 0.2 e/Å*^{3} *(Sn atom is at the origin)*

Figure 3.36 MEM 2D electron density map of $Sn_{0.75}Ge_{0.25}Te$ and $Sn_{0\text{-}88}Ge_{0.12}Te$ on (110) plane, Contour range is (0-15) e/Å^3 with step size is 0.2 e/Å^3 (Sn atom is at the origin)

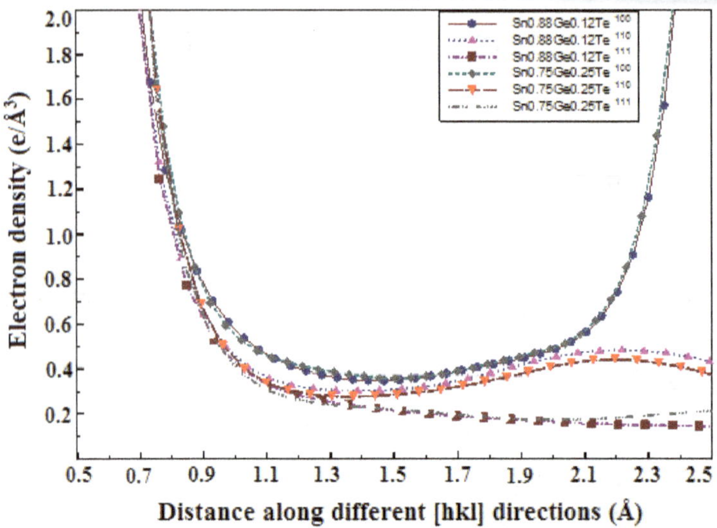

Figure 3.37 One dimensional low MEM electron density profiles of $Sn_{0.75}Ge_{0.25}Te$ and $Sn_{0.88}Ge_{0.12}Te$ along [100], [110] and [111] (Sn atom is at the origin)

3.5.6.2 DISCUSSION OF THE RESULTS

From the MEM refinement process, it is found that the number of refinement cycles is slightly higher for $Sn_{0.75}Ge_{0.25}Te$ and $Sn_{0.88}Ge_{0.12}Te$, probably due to the higher number of electrons in the unit cell, having the number of pixels being the same in both systems. The R_{MEM} and wR_{MEM} values are also very low for the two systems as seen in table 3.11.

The three dimensional electron densities imposed on the structure of $Sn_{0.75}Ge_{0.25}Te$ and $Sn_{0.88}Ge_{0.12}Te$ are shown in figure 3.32. It is found from figure 3.32 that the electron density clouds around the Sn atoms are enlarged in $Sn_{0.88}Ge_{0.12}Te$, which perfectly implies the presence of a higher number of electrons in $Sn_{0.88}Ge_{0.12}Te$ compared to $Sn_{0.75}Ge_{0.25}Te$. The 2D electron density imposed in the 3D structure on the (100) and the (110) planes are visualized in figures 3.33 and 3.34 respectively, which give one a better clarity of their interactions.

The 2D MEM electron density maps have been constructed on the (100) plane of both $Sn_{0.75}Ge_{0.25}Te$ and $Sn_{0.88}Ge_{0.12}Te$ and are presented in figure 3.35 considering the same contour range and contour interval for comparison. The maps on the (100) planes can be visualized and interpreted considering the unit cell structure of SnTe. Comparing these two figures, it is found that there is meager expansion of the atomic core of the Sn atom and an additional electron density contour is visible in the case of $Sn_{0.88}Ge_{0.12}Te$, which may be due to a greater number of electrons in $Sn_{0.88}Ge_{0.12}Te$ which has been distributed as extra contour lines. Similar trends are also visible in the (110) plane as shown in figure 3.36. Both the figures 3.35 and 3.36 explicitly show the covalent type of bonding between the atoms Sn(Ge) and Te, the co-valency being judged by the strengths of the bonding, which are 0.346 e/Å^3 and 0.359 e/Å^3 for x=0.12 and 0.25 respectively.

The one-dimensional electron densities are shown in figure 3.37 for $Sn_{0.75}Ge_{0.25}Te$ and $Sn_{0.88}Ge_{0.12}Te$ along the [100], the [110] and the [111] directions. The positions of minimum electron densities and the density values are given in table 3.12. Along the bonding direction in $Sn_{0.88}Ge_{0.12}Te$ and $Sn_{0.75}Ge_{0.25}Te$ *i.e.*, the [100] direction, the mid-bond density is found to be 0.3459 e/Å3 and 0.3585 e/Å^3 and at a distance of 1.4695 Å and 1.5525 Å respectively. The slightly reduced electron density in $Sn_{0.88}Ge_{0.12}Te$ can be due to the expansion of the cell.

3.5.7 INDIUM ANTIMONIDE (InSb)

MEM formalism, an effective tool for the analysis of crystalline materials has being used for the analysis of the electron density distribution inside the single crystal InSb. It produces the so-called "super resolution" electron density distributions [Saravanan, 2009]. The MEM analysis of InSb may give a better knowledge of the bonding nature between the individual atoms and electron distribution around the indium and the antimony atoms, which is highly necessary for the design of better thermoelectric materials.

In this work on InSb, the MEM refinements were carried out by dividing the unit cell into 128x128x128 pixels. The initial electron density at each pixel is fixed uniformly as F_{000}/a_0^3 = 1.478 e/Å^3, where F_{000} is the total number of electrons in the unit cell and a_0 is the cell parameter. The Lagrangian parameter is suitably chosen so that the convergence criterion C=1 is reached after a minimum number of iterations. For the numerical MEM computations, the software package PRIMA [Izumi and Dilanian, 2002] was used. For the 1D, 2D and 3D representation of the electron densities, the program VESTA [Momma and Izumi, 2006] package was used.

3.5.7.1 MEM REFINEMENTS AND RESULTS

The MEM parameters are given in table 3.13. The 3D electron density distributions in the form of iso-surface in the unit cell are represented in figure 3.38 for the analysis with and without h+k+l =4n+2 reflections. The 2D electron density distribution for the reflections with and without h+k+l =4n+2 on the (110) plane are given in figure 3.39. Similarly, the 2D density distributions on the (100) plane are shown in figures 3.40. The one-dimensional electron density profiles for the analysis of reflections with and without h+k+l =4n+2 reflections, along the [100], the [110] and the [111] directions are represented in figures 3.41 and 3.42 respectively. Tables 3.14 and 3.15 give the numerical values of the electron densities along the three crystallographic directions [100], [110] and [111].

3.5.7.2 DISCUSSION OF THE RESULTS

The 3D electron density distribution of InSb as seen in figure 3.38 shows the bonding between In and Sb atoms and the interaction of the electron clouds of these atoms. Though there are minute differences in the electron density distributions when the quasi-forbidden reflections are included or excluded from the analysis, the interactions of the In and the Sb atoms are the same in general, in both cases.

The two dimensional electron density distributions on the (110) plane with h+k+l=4n+2 and without h+k+l=4n+2 type reflections are shown in figure 3.39, prove a strong covalent interaction between the In and the Sb atoms. Since InSb belongs to a FCC structure, the two dimensional electron density distribution on the (100) plane, both with h+k+l=4n+2 and without h+k+l=4n+2 type reflections shown in figure 3.40 show the distribution of the In atoms only. When the h+k+l=4n+2 type reflections are excluded, the electron clouds show symmetric and uniformly distributed patterns both in the (100) and the (110) planes.

The one-dimensional electron density profiles along the (100), the (110) and the (111) directions of the unit cell of InSb with h+k+l=4n+2 and without h+k+l=4n+2 type reflections are shown in figures 3.41 and 3.42 for the low density region. The one-dimensional electron density values along the bonding direction [111] with h+k+l=4n+2 and without h+k+l=4n+2 type reflections reveal a mid bond electron density of 0.2564 e/Å^3 and 0.2632 e/Å^3 respectively, between In and Sb atoms at a distance of 1.399Å from the origin. This indicates the covalent nature of InSb and proves that InSb is a very good thermoelectric material due to heavy charge concentration but with a localized nature. The 2D electron density distributions also support the localized charge distribution in InSb.

Table 3.13 The parameters used and obtained in MEM refinements of InSb

Parameter	with h+k+l=4n+2	without h+k+l=4n+2
Number of cycles	1064	5160
Lagrange parameter (λ)	0.022	0.006
Prior electron density (e/Å^3)	1.478	1.478
R_{MEM} (%)	1.9	1.6
wR_{MEM} (%)	2.3	2.0
Resolution (Å)	0.05	0.05

Table 3.14 The 1D electron density of InSb (with h+k+l=4n+2 type reflections) along the three directions in the unit cell

Direction	Position (Å)	Electron density (e/Å^3)
[100]	0.000	894.5500 (In peak)
[100]	3.232	0.0136
[100]	6.463	894.5500
[110]	2.285	0.1949
[111]	1.399	0.2564 (Mid-bond)
[111]	2.798	1289.7100 (Sb peak)

Table 3.15 The 1D electron density of InSb (without h+k+l=4n+2 type reflections) along the three directions in the unit cell

Direction	Position (Å)	Electron density (e/Å^3)
[100]	0.000	981.8100 (In peak)
[100]	3.231	0.0137
[100]	6.463	981.8100
[110]	2.285	0.2377
[111]	1.399	0.2632 (Mid-bond)
[111]	2.788	1182.9000 (Sb peak)

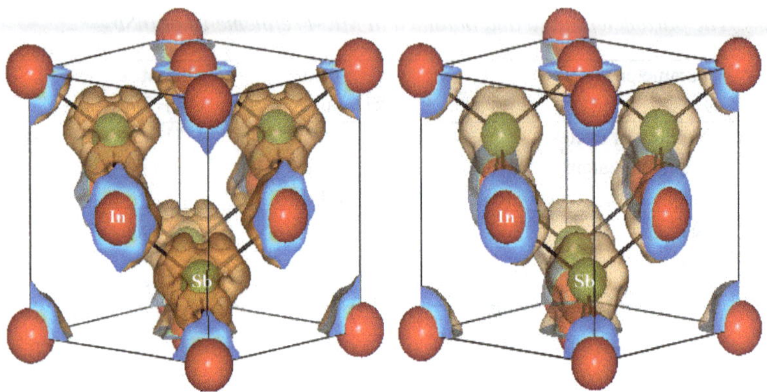

Figure 3.38 3D iso-surface of the electron density of InSb in the unit cell (analysis with and without h+k+l=4n+2 reflections, respectively)

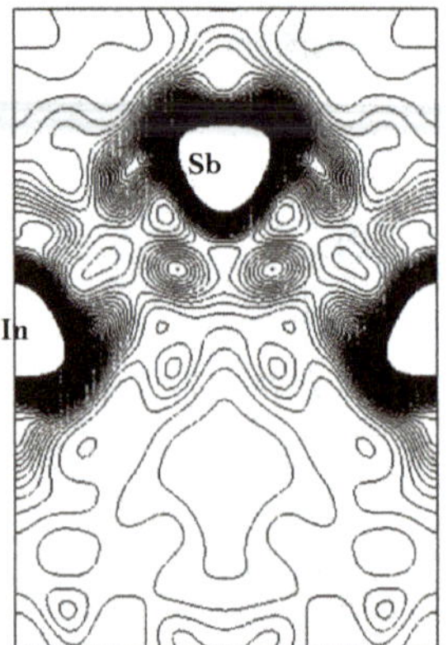

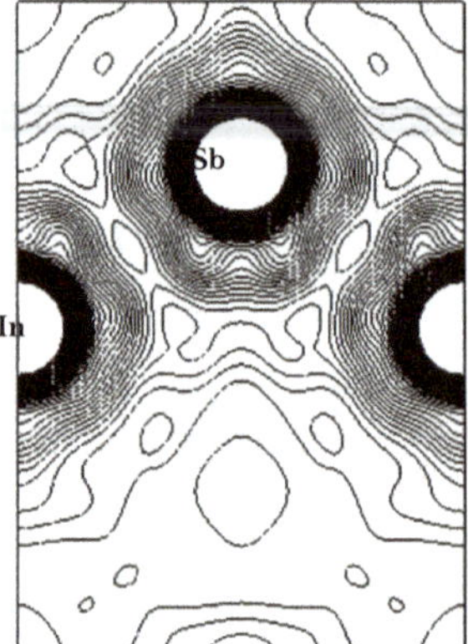

Figure 3.39 MEM electron density contour map of InSb on (110) plane (analysis with and without h+k+l=4n+2 reflections, respectively)

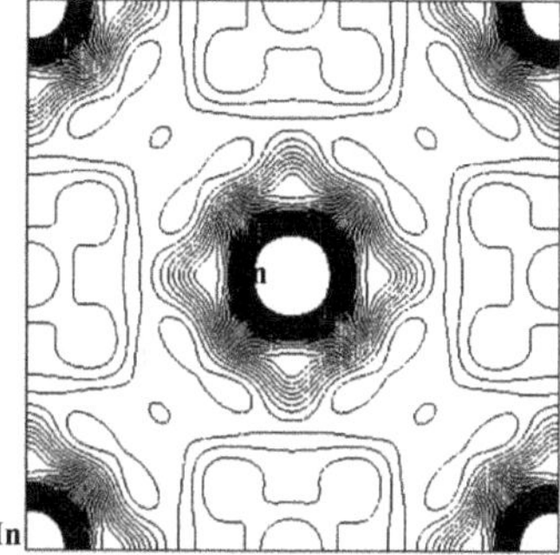

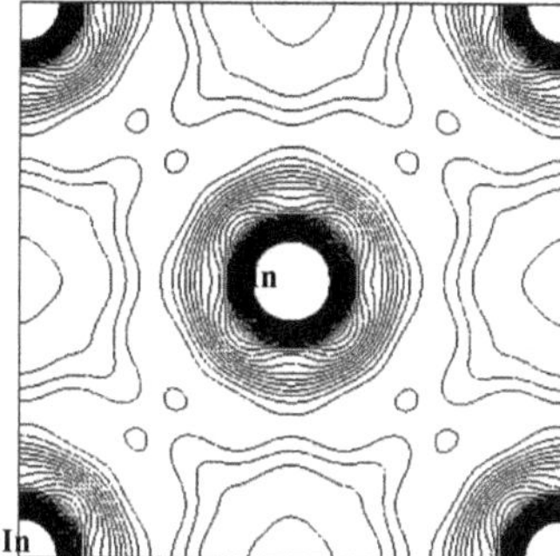

Figure 3.40 MEM electron density map of InSb on (100) plane (analysis with and without h+k+l=4n+2 reflections, respectively)

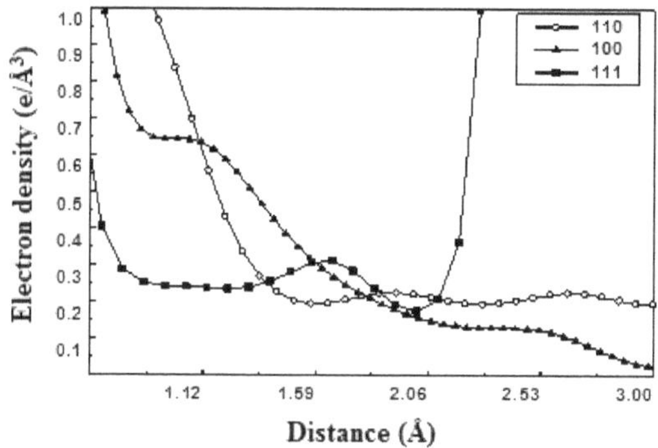

Figure 3.41 One dimensional MEM electron density profiles of InSb along [100], [110] and [111] directions (analysis with h+k+l=4n+2 reflections)

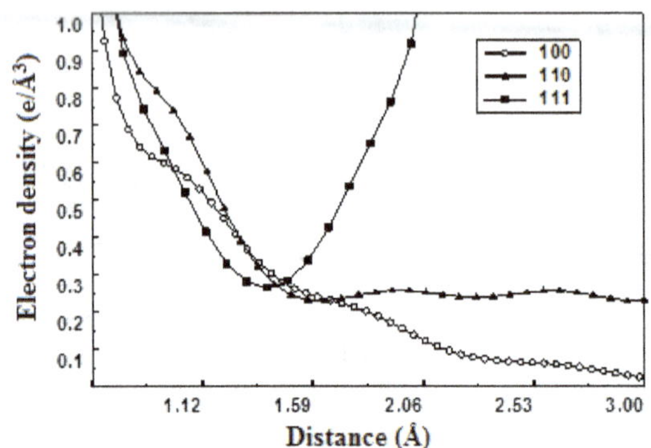

Figure 3.42 One dimensional MEM electron density profiles of InSb along [100], [110] and [111] directions (analysis without h+k+l=4n+2 reflections)

REFERENCES

[1] Bricogne G, Acta Cryst. Vol. A44 (1988) p. 517.

[2] Bricogne G, Gilmore C.J, Acta Cryst. Vol. A46 (1990) pp. 284-297.

[3] Bricogne G, Gilmore C. J, Acta Cryst. Vol. A46 (1990) p. 284.

[4] Bricogne G, Acta Cryst. Vol. D49 (1993) p.37.

[5] Burger K, Prandl W, Acta Cryst. Vol. A55 (1999) p.719. *http://dx.doi.org/10.1107/S0108767399001191*

[6] Brown I.D, The Chemical Bond in Inorganic Chemistry (2002).

[7] Bruning H, PhD Thesis, University of Twente, Enschede, Netherlands (1992).

[8] Collins D, Nature Vol. 298 (1982) p. 49-51. *http://dx.doi.org/10.1038/298049a0*

[9] Cox D.E, Papoular R, Mat. Sci. Forum (1996) pp. 228-233.

[10] Craven B.M, Weber H.P, He X, Tech. Report TR-87-2, Department of Crystallography, University of Pittsburgh, Pittsburgh,Pa. (1987).

[11] Gilmore Chr J, Acta. Cryst. Vol. A52 (1996) pp. 561-589. *http://dx.doi.org/10.1107/S0108767396001560*

[12] Graafsma H, De Vries R. Y. J, Appl. Cryst. Vol.32 (1999) p. 683. De Vries R.Y, Brils W.J, Feil D, Acta Crystallogr. Vol. A50 (1994) p. 383.

[13] Hansen N.K, Coppens P, Acta Crystallogr. Sect. A: Cryst. Phys. Diffr. Theor. Gen. Crystallogr. Vol. A34 (1978) p. 909.
http://dx.doi.org/10.1107/S0567739478001886

[14] Hirshfeld F.L, Acta Crystallogr. Sect. B: Struct. Crystallogr. Cryst. Chem. Vol. B27 (1971) p. 769.
http://dx.doi.org/10.1107/S0567740871002905

[15] Hirshfeld F.L, Isr. J. Chem. Vol.16 (1977) p. 226.
http://dx.doi.org/10.1002/ijch.197700037

[16] Isoda Y,Nagai T, Fuziu H, Imai Y, Shinohara Y, Proceedings of ICT'06, IEEE (2006) p. 406.

[17] Isoda Y,Nagai T, Fuziu H, Imai Y, Shinohara Y, Proceedings of ICT'07, IEEE (2008) p. 251.

[18] Izumi F, Dilanian R.A, Recent Research Developments in Physics Vol. 3, Part II, ed. by S. G. Pandalai, Transworld Research Network, Trivandrum (2002) p. 699-726. (ISBN 81-7895-046-4).

[19] Marks L. D, Landree E, Acta Cryst. Vol. A54 (1998) p. 296.
http://dx.doi.org/10.1107/S0108767397016917

[20] Momma K, Izumi F, Commission on Crystallogr. Comput IUCr Newslett. Vol. 7 (2006) p. 106.

[21] Momma K, Izumi F, Super-fast Program PRIMA for the Maximum-Entropy Method, J. Appl. Cryst. Vol. 44 (2011) pp. 1272-1276.
http://dx.doi.org/10.1107/S0021889811038970

[22] Namiki, Tsukuba, Ibaraki, Advanced materials Laboratory, National institute for materials science, 1-1 Japan, Vol. 305 (2004) p. 44.

[23] Papoular R, Cox D. E, Europhys. Lett. Vol.32 (1995) pp. 337-342.
http://dx.doi.org/10.1209/0295-5075/32/4/009

[24] Rietveld H. M., J. Appl. Cryst. Vol. 2 (1969) pp. 65-71.
http://dx.doi.org/10.1107/S0021889869006558

[25] Saravanan R, Physica Scripta (2009).

[26] Sakata M, Sato M, Acta Cryst. Vol. A46 (1990) p. 263
http://dx.doi.org/10.1107/S0108767389012377

[27] Sakata M T, Uno M, Takata, Mori R, Acta Crystallogr. Sect. A: Found. Crystallogr. Vol. A39 (1992) p. 47.

[28] Sivia D. S, David W. I. F, Acta Cryst. vol. A50 (1994) p. 703. *http://dx.doi.org/10.1107/S0108767394003235*

[29] Takata M, Sakata M, Kumazawa S, Larsen F, Iversen B, Acta. Cryst. Vol. A50, (1994) p. 330. *http://dx.doi.org/10.1107/S0108767393011523*

[30] Takata M, Sakata M, Acta Cryst. vol. A52 (1996) p. 287. *http://dx.doi.org/10.1107/S0108767395014917*

[31] Wilkins S.W, Varghese J.N, Lehmann M. S, Acta Cryst. Vol. A39 (1983) p. 47. *http://dx.doi.org/10.1107/S0108767383000082*

[32] Zhang Q, He J, Zhu T.J, Zhang S.N, Zhao X.B, Tritt T.M, Appl. Phys. Lett. Vol. 93 (2008) pp. 102-109.

CHAPTER IV

Results and Discussion Based on Pair Distribution Function (PDF)

Abstract

The study of local structure i.e., the inter atomic ordering is of great interest, because it provides information of the first, second and third coordination shells, in terms of the number of electrons in the corresponding shell as well as the distance between the shells. The first method of choice to study inter atomic ordering is the X-ray absorption fine structure method (XAFS) [Stumm, 1989; Rehr and Albers, 2000; Filipponi *et al.*, 1995; deGroot, 2001; Lytle, 1999; Sayers *et al.,* 1971]. The alternate method for studying a local structure is the powder X-ray method or the neutron diffraction data method, i.e., pair distribution function (PDF). It is a theoretical approach which results from a Fourier transform of the powder diffraction spectrum containing both Bragg peaks and diffuse scattering into real space [Egami and Billinge, 2003]. The pair distribution function (PDF), that can be obtained by Fourier transform of powder diffraction data, traditionally has been used to describe short-range correlations in atomic positions. In recent years, this technique has been developed further and allows one to achieve an extremely good agreement between calculated and experimental PDFs for crystalline materials [Chung, 1997; Thorpe *et al.,* 1998, 2002]. Hence, in this work, an attempt has been made to study the short range inter atomic correlations using powder X-ray diffraction data. It was believed and proven that the high Q (momentum transfer) synchrotron data is essential for the short range atomic correlation study [Egami and Billinge, 2003]. Though the Q value of the powder X-ray data sets (used in this work) is small compared to the neutron or synchrotron data, an attempt has been successfully made for the study of atomic pair distribution of the following thermoelectric materials: magnesium silicide (Mg_2Si), lead telluride (PbTe), bismuth telluride (Bi_2Te_3) and antimony telluride (Sb_2Te_3).

Keywords

Pair Distribution Function, PDF Achievement, PDFgetX, PDFFIT, Lead Telluride, Antimony Telluride, Bismuth Telluride, Magnesium Silicide

Contents

4.1 PDF ACHIEVEMENT PROCESS

The experimental PDF is obtained from the coherently scattered intensities which are extracted from X-ray diffraction pattern by applying appropriate correction for flux, background, Compton scattering and sample absorption. The intensity is normalized in an absolute electron unit and reduced to a structure function. The PDF is obtained by taking the Fourier transform of the reduced structure function.

PDFgetX [Jeong *et al.*, 2001] is a program used to obtain the observed atomic pair distribution function (PDF) from X-ray powder diffraction data. The raw powder X-ray data is obtained using the Cu-Kα target for analytical processing. PDFgetX [Jeong *et al.*, 2001] can reduce the raw data into a more convenient format from which the pair distribution function can be obtained for further analysis. PDF is the instantaneous atomic number density - density correlation function which describes the atomic arrangement in materials. A useful characteristic of the PDF method is that it gives both local and average structural information because both Bragg peaks and diffuse scattering are used in the analysis. Also, from the PDF peak width, it is possible to obtain the information of bond length distribution for static, thermal and correlated atomic thermal motion [Petkov *et al.*, 1999]. By contrast, an analysis of the Bragg scattered intensities alone, by a Rietveld type [Rietveld, 1969] analysis for instance, yields the average crystal structure only and the extended X-ray absorption fine structure (EXAFS) gives nearest-neighbor and next nearest-neighbor distance information. The PDF method has found more application in the study of local structural disorder in crystalline materials, where some deviation from the average structure is expected to take place.

Obtaining total scattering structure function and pair distribution function from raw diffraction data requires many corrections for experimental effects such as absorption, polarization corrections and removing of Compton and multiple scattering contributions to the elastic scattering. In the data analysis procedure, PDFgetX [Jeong *et al.*, 2001] is composed of four main blocks, preliminary data reduction, building PDFgetX [Jeong *et al.*, 2001] input file, refine structure function and pair distribution function calculation. Also, it needs proper error propagation to be used in modeling of PDF using PDFfit [Proffen and Billinge, 1999] to yield structural parameters.

In this work, to get the observed and calculated PDF, the graphical software PDFgui [Farrow *et al.*, 2007] was used, which is a graphical environment for PDF fitting. This allows powerful usability features such as real time plotting and remote execution of the fitting program and visualizing the results locally.

During the process of PDF fitting various structural factors like lattice parameters, phase scale factor, linear atomic correlation factor, quadratic atomic correlation factor, spherical nano particle amplitude correction, low r peak sharpening, peak sharpening cutoff and cutoff for profile setup functioning were refined. The data configuration parameters were the PDFfit [Proffen and Billinge, 1999] range from step size, data scale factor, upper limit for Fourier transform to obtain data PDF, resolution damping factor, resolution peak broadening factor, data collection temperature and doping concentration levels etc., which can be refined to get an accurate PDF fitting. Finally, the observed and calculated PDF's are visualized and compared. In the present work, the local structures of the following materials have been characterized using the pair distribution function: 1. Mg_2Si, 2. PbTe, 3. Bi_2Te_3 and 4. Sb_2Te_3.

4.2 MAGNESIUM SILICIDE (Mg_2Si)

In order to understand the local structure of Mg_2Si using the real space analysis of the X-ray powder data, the observed pair distribution function was obtained using the above mentioned PDF achievement process using software packages PDFgetX [Jeong *et al.*, 2001] and PDFgui [Farrow *et al.*, 2007]. The refined parameters of Mg_2Si using pair distribution function are given in table 4.1. The fitted PDF's are given in figure 4.1. It is worthy to point out that the matching between the observed and calculated PDF's is perfect with a small error function. The refined cell constant from PDF analysis is 6.3568(1) Å which is slightly different from the reported, 6.39 Å [Wyckoff, 1963]. The nearest neighbor atom distances obtained from the PDF analysis have been given in table 4.2. Figure 4.1 shows PDF peaks representing the interaction between Si-Si, Mg-Mg and Si-Mg atomic pairs. PDF peaks up to a distance of 13 Å are suitably indexed as shown in figure 4.1. The PDF peak at 2.94 Å corresponds to the Si-Mg bond length along [111] direction. The mid-bond position turns out to be half of the bond length, i.e., (2.94/2 Å), which is 1.47 Å. This position is almost the same as the mid-bond position along [111] direction (1.475 Å), obtained from the MEM analysis. Similarly, from the MEM analysis along [110] direction, a small hump is seen at 2.251 Å. From the PDF analysis, the Si-Si bond length turns out to be 4.36 Å. Thus the Si-Si mid bond position is 2.18 Å, which compares with the hump seen in MEM analysis at 2.251 Å. Thus all the PDF peaks can be suitably indexed. Hence, the MEM technique combined with PDF analysis can give much more information of the electron density and local structure of the system under consideration.

Table 4.1 The refined parameters from PDF of Mg_2Si

Refined parameters	Values
Data range (Å)	0.02-30.0
Refinement range (Å)	1.7-20.0
Q_{max} ($Å^{-1}$)	6.7

Table 4.2 Nearest neighbour distances from PDF analysis of Mg_2Si

Atom pair	Inter-atomic distances from PDF analysis (Å)		Calculated* inter-atomic distances(Å)
	Observed	Calculated	
Si-Mg	2.94	2.88	2.75
	5.36	5.32	5.27
	6.90	7.00	6.92
	8.06	8.10	8.25
	9.38	9.36	9.39
	10.26	10.26	10.41
	11.98	11.98	11.88
Mg-Mg; Si-Si	4.36	4.40	4.49
Mg-Mg; Si-Si	13.38	13.40	13.47

* [Laugier and Bochu, 2002]

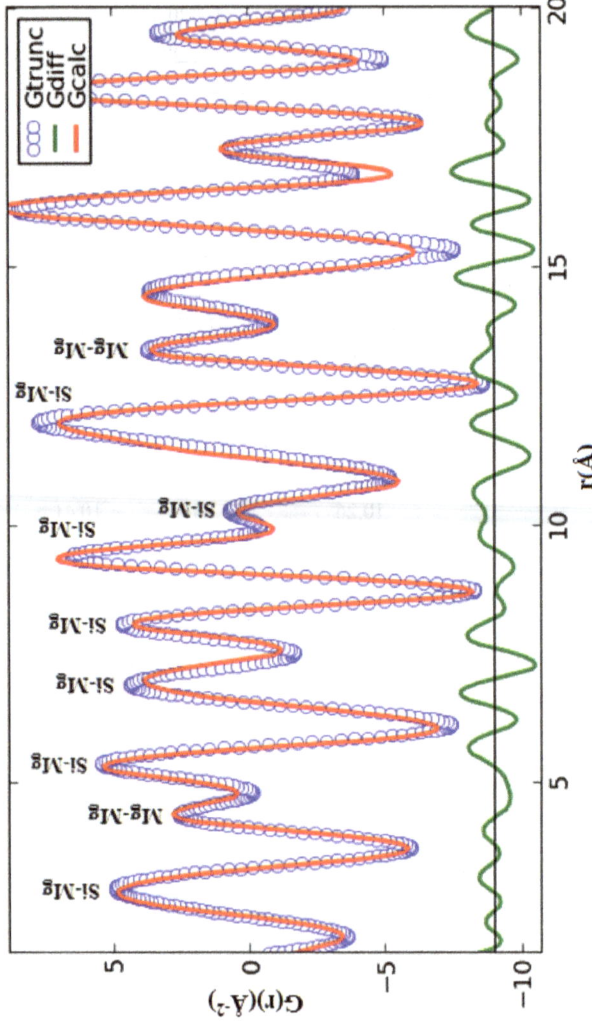

Figure 4.1 The fitted observed and calculated PDF's of Mg_2Si

4.3 LEAD TELLURIDE (PbTe)

In order to understand the local structure of PbTe using the real space analysis of the X-ray powder data, the observed pair distribution function was obtained using the software package PDFgetX [Jeong *et al.*, 2001]. The observed PDF was fitted with the calculated one using the software PDFgui [Farrow *et al.*, 2007]. The fitted PDF's are given in figure 4.3. The refined cell constant from PDF analysis is 6.4473 (0.0002) Å which is comparable to the reported value 6.454 Å [Wyckoff, 1963]. Some of the nearest neighbor atom distances obtained from the PDF analysis have been given in table 4.4.

Since, PDF analysis relies on the local structure, the local undulations in the bond charges and the interactions between atoms lead to deviations in the cell parameter from that obtained only from the average Bragg positions. Since the deviation in the cell constant in PDF analysis is very small from that reported, it is obvious that there are not many local distortions in the structure of the pure PbTe, except those resulting from bonding interactions between atoms.

Figure 4.2 shows a small peak at 2.22 Å from the origin corresponding to the interaction discussed above. Since, PDF analysis deals with atomic number density, all the interactions between Te and Pb atoms along [110] direction comes out as a peak in the PDF analysis. Other PDF peaks up to a distance of 13 Å are suitably indexed as shown in figure 4.2. The PDF peak at 3.26 Å corresponds to the Te-Pb distance along the [100] direction. The minimum electron density along the [100] direction occurs at 1.618 Å as seen from the MEM analysis. In the present PDF analysis, the mid bond position happens at 1.63 Å (3.26/2 Å). Hence, the Te atom has a slightly smaller atomic radius than the Pb atom, since the MEM electron density minimum occurs at 1.618 Å and not at 1.63 Å. From the PDF bond length, one can assign the experimental radii of Te and Pb atoms as 1.618 Å and 1.642 Å respectively. The doublets at 7.22 Å and 7.72 Å correspond to the Te-Pb and Pb-Pb atomic distances along different directions. Hence, the MEM technique combined with PDF analysis can give much more information on the electron density and local structure of the system under consideration.

Table 4.3 The refined parameters from PDF of PbTe

Refined parameters	Values
Data range (Å)	0.02-30.0
Refinement range (Å)	2.0-20.0
Q_{max} ($Å^{-1}$)	6.75

Table 4.4 Nearest neighbour distances from PDF analysis of PbTe

Atom pair	Inter-atomic distances from PDF analysis (Å)		Calculated inter-atomic distance (Å) [Laugier and Bochu, 2002]
	Observed	Calculated	
Te-Te	2.24	2.20	2.28
Te-Pb	3.26	3.26	3.22
Te-Te	4.46	4.48	4.56
Te-Pb	5.68	5.64	5.58
Te-Pb	7.22	7.22	7.21
Pb-Pb	7.72	7.00	7.89
Te-Pb	9.92	9.86	10.19
Te-Pb	11.96	11.98	11.62
Te-Pb	13.40	13.42	13.29

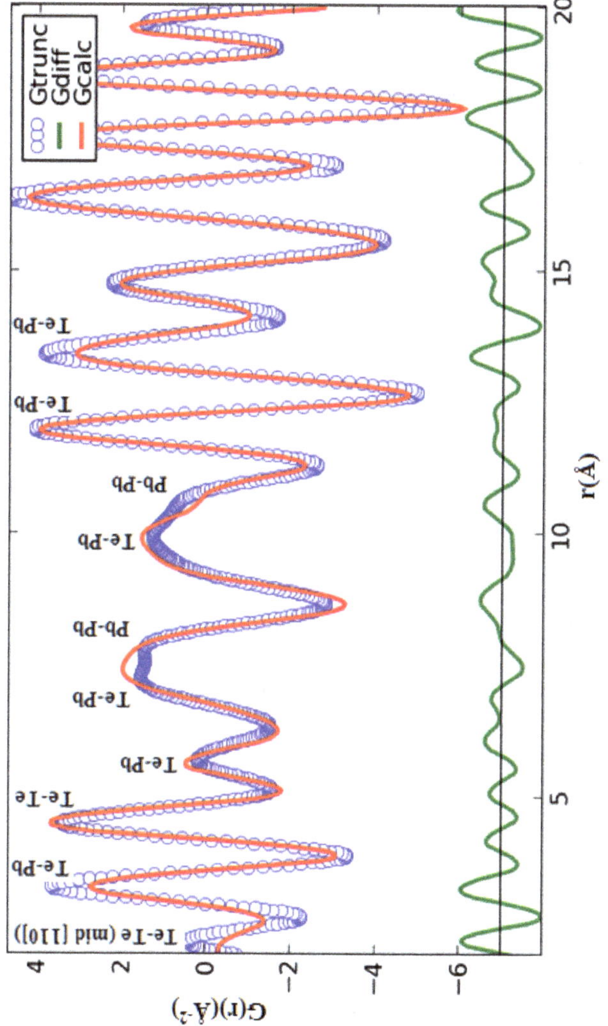

Figure 4.2 The fitted observed and calculated PDF's of PbTe

4.4. BISMUTH TELLURIDE (Bi_2Te_3)

The powder X-ray intensity data of Bi_2Te_3 has been utilized for the analysis of the pair distribution function (PDF). The observed PDF's have been obtained using the software package PDFgetX [Jeong *et al.*, 2001]. The PDF for Bi_2Te_3 was refined using PDFgui [Farrow *et al.*, 2007] and the comparison of observed and calculated PDF's has been made and analyzed.

The refined pair distribution functions (the Fourier transform of $S(\vec{Q})$, the reduced structure factor) are given in figure 4.3. An excellent matching between the observed and the calculated PDF's can be noticed. The results of the PDF analysis have been given in table 4.5. The first three nearest neighbour distances along with the bonding atoms are given in table 4.6. The PDF refinement can be considered as equivalent to matching the observed X-ray powder data with a model with too little structure parameters. Moreover, high Q data is required for these types of analyses on a short range order and local structure of materials [Jeong *et al.*, 2001; Egami, 1990]. In spite of these factors, an attempt has been made to use the X-ray powder data for this analysis and the results obtained were reasonable. The inter-atomic distances were calculated using the software GRETEP [Laugier and Bochu, 2002, http://www.ccp14.ac.uk/tutorial/imgp/].

The first nearest neighbour is observed at a distance of r = 3.003 Å in the PDF table 4.6. In that distance Bi-Te atoms are located with an inter-atomic distance of r = 3.196 Å. In the same region, there are possibilities of the location of Bi-Te1 atoms with an inter-atomic distance of r = 3.124 Å. The second nearest neighbour is found to be at a distance of r = 6.6406 Å. In that particular distance, one can locate Bi-Bi atoms at an inter-atomic distance of r = 6.392 Å. There are possible locations of Bi-Te, Bi-Te1 and Te-Te1 with the inter-atomic distance of 6.987 Å, 6.954 Å and 6.415 Å respectively. Similarly, one can find the third and fourth nearest neighbours.

Table 4.5 The refined parameters from PDF of Bi_2Te_3

Refined parameters	Values
Data range (Å)	0.02-30.0
Refinement range (Å)	2.4-28.4
Q_{max} ($Å^{-1}$)	7.00

Table 4.6 Nearest neighbour inter-atomic distance of Bi_2Te_3

Nearest neighbour	Distance observed r(Å) [Laugier and Bochu, 2002]	Bi- Bi (Å)	Bi-Te (Å)	Te - Te1 (Å)	Bi-Te1 (Å)
1	3.003		3.196		3.124
2	6.6406	6.392	6.987	6.415	6.954
3	7.59	7.676			7.206

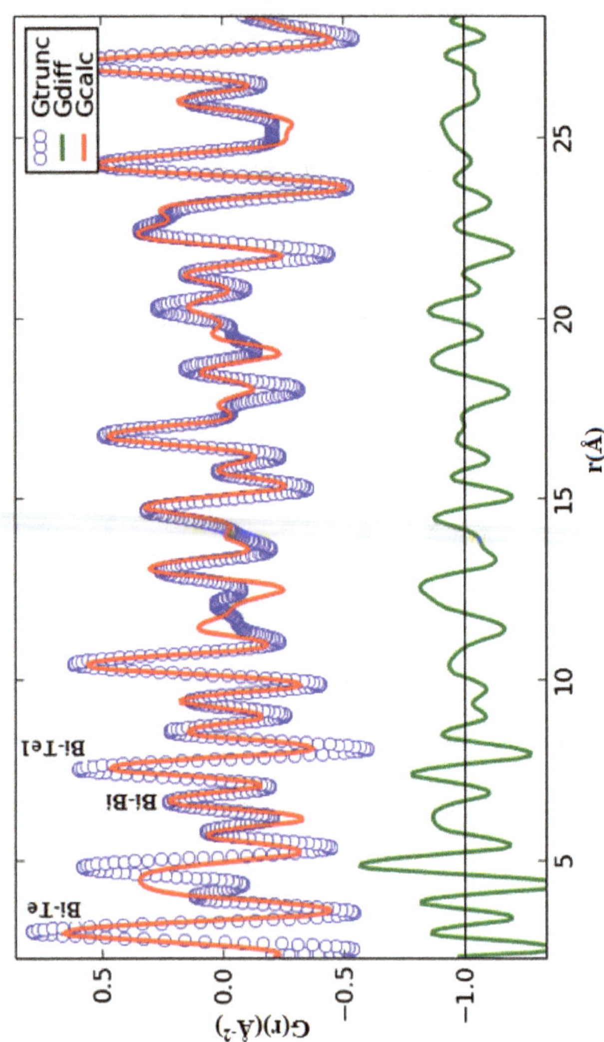

Figure 4.3 The fitted observed and calculated PDF's of Bi_2Te_3

4.5 ANTIMONY TELLURIDE (Sb_2Te_3)

The local structure of Sb_2Te_3 has been analyzed in terms of bond length, concentration of atoms at a particular distance etc. By using the software package PDFFIT [Farrow *et al.*, 2007] the pair distribution function can be refined leading to various structural parameters like thermal vibrations of atoms, bond lengths, occupancy, atomic concentration etc. The determination of average structure based on powder diffraction data routinely done using Rietveld method [Rietveld, 1969], is very similar to the full profile refinement of the atomic pair distribution function. The analysis of Bragg scattering assumes a perfect long-range order of the crystal. The observed raw powder intensities of Sb_2Te_3 are used for getting the observed PDF's using the software package PDFgetX [Jeong *et al.*, 2001] along with multiple scattering corrections.

In the present work, the powder X-ray data of Sb_2Te_3 has been utilized for the analysis of pair distribution function. The observed PDF's have been obtained from the software package PDFgetX [Jeong *et al.*, 2001] as mentioned earlier. PDFgetX can help reduce the raw data into a more convenient format from which the analysis can be done within a range of r = 20 Å with a step size of 0.02 Å. The upper limit of the Fourier transform of the data set is Q_{max}=7Å^{-1}. After the preliminary data reduction, we have used PDFgui [Farrow *et al.*, 2007] for fitting the theoretical structural models with the experimental PDF in a refinement procedure.

Structural parameters like lattice parameter, phase scale factor, linear and quadratic atomic correlation factors along with thermal amplitudes and occupancy of the atoms are refined. In the end, the observed and calculated PDF's are obtained. The parameters in the PDF refinement and results were shown in the table 4.7. Figure 4.4 shows the observed and calculated pair distribution functions for Sb_2Te_3.The magnitudes of the distances were calculated. The inter-atomic distances were calculated by means of GRETEP software [Laugier and Bochu, 2002, http://www.ccp14.ac.uk/tutorial/imgp/]. Then, the calculated distance from PDFgui [Farrow *et al.*, 2007] was compared with the experimental results and the presence of atoms like Sb, Te were observed and their inter-atomic distances were calculated and shown in table 4.8.

The first nearest neighbor is observed at a distance of r = 3.079 Å in the PDF table 4.8. In that distance Sb-Te atoms are located with an inter-atomic distance of 3.107 Å. In the same region, there are possibilities of the location of Sb–Te1 atoms with a inter-atomic distance of r = 3.043 Å. The second nearest neighbour is at a distance of r = 4.336 Å. In that particular distance one can locate Sb-Sb atoms at a inter-atomic distance of r = 4.264

Å. There are possible locations of Sb-Te and Sb-Te1 with the inter-atomic distance of 4.264 Å and 5.238 Å respectively.

Table 4.7 The refined parameters from PDF of Sb_2Te_3

Refined parameters	Values
Data range(Å)	0.02-30.0
Refinement range(Å)	2.0-30.0
Q_{max}(Å^{-1})	7.00

Table 4.8 Inter-atomic distance between atoms of Sb_2Te_3

Nearest neighbour	Distance observed r(Å) [Laugier and Bochu, 2002]	Sb-Sb (Å)	Sb-Te (Å)	Te-Te1 (Å)	Sb-Te1 (Å)
1	3.079		3.107		3.043
2	4.336	4.264		4.264	5.238
3	6.653	6.364	6.784	6.473	
4	8.85	7.661			7.021

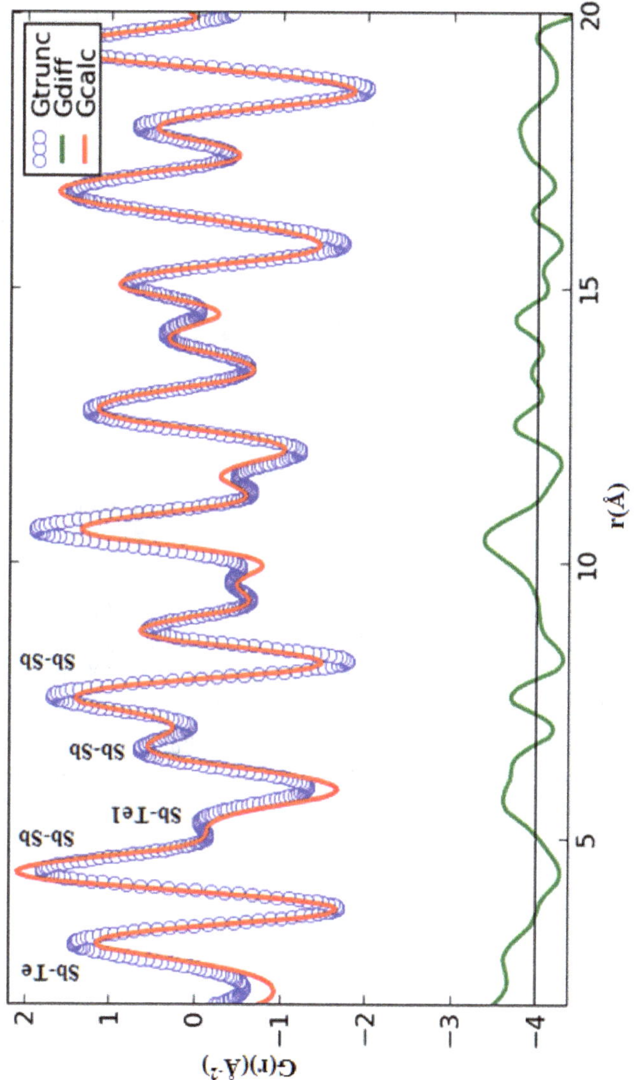

Figure 4.4 *The fitted observed and calculated PDF's of* Sb_2Te_3

REFERENCES

[1] Billinge S.J.L, Thorpe M.F, Advances in Pair Distribution Profile Fitting in Alloys in Local Structure from Diffraction, Edited By Plenum Press, New York (1998) pp. 157-174.

[2] Chung J.S, Thorpe M.F, Local atomic structure of semiconductor alloys using pair distribution function Phys. Rev. Vol. B 59 (1997) pp. 1545-1553.
http://dx.doi.org/10.1103/PhysRevB.55.1545

[3] de Groot F, High-resolution X-ray emission and X-ray absorption spectroscopy. Chemical Reviews, Vol. 101 (2001) pp. 1779–1808.
http://dx.doi.org/10.1021/cr9900681

[4] Egami T, Billinge S.J.L, Underneath the Bragg Peaks: Structural Analysis of Complex Material, Oxford University Press, London (2003).

[5] Egami T, JIM. 31, No.3 (1990) p.163.
http://dx.doi.org/10.2320/matertrans1989.31.163

[6] Filipponi A. Di Cicco A, C.R. Natoli C.R., X-ray absorption spectroscopy and n-body distribution functions in condensed matter. Physical Review B Vol. 52/21 (1995) pp. 15122–15148.
http://dx.doi.org/10.1103/PhysRevB.52.15122

[7] Farrow C.L, Juhás P, Liu J.W, Bryndin D, Bozin E.S, Bloch J, Proffen Th, Billinge S.J.L,"PDFfit2 and PDFgui: Computer programs for studying nanostructure in crystals", J. Phys. Condens. Matter Vol. 19 (2007) p. 335219.
http://dx.doi.org/10.1088/0953-8984/19/33/335219

[8] Jeong I.K, Thompson J, Proffen Th, Perez A, Billinge S.J.L,, "PDFGetX, A program for obtaining the Atomic Pair Distribution Function from X-ray Powder Diffraction Data" (2001).

[9] Laugier et B Bochu J, GRETEP, Domaine universitaire BP 46, 38402 saint martin D'Heres
http: /www.inpg.fr / LMGP, (2002)

[10] Lytle F.W, The EXAFS family tree: a personal history of the development of extended X-ray absorption fine structure, Journal of Synchrotron Radiation Vol. 6 (1999) pp. 123–134.
http://dx.doi.org/10.1107/S0909049599001260

[11] Petkov V, Jeong I.K, Chung J.S, Thorpe M.F, Kycia S, Billinge S.J.L, 'High-real space resolution measurement of the local structure of InxGa 1-xAs semiconductor alloys using X-ray diffraction', Phys. Rev. Lett. 83 (1999) p. 4089. *http://dx.doi.org/10.1103/PhysRevLett.83.4089*

[12] Proffen T, Billinge S.J.L, J. Appl. Cryst. Vol. 32 (1999) p.572. *http://dx.doi.org/10.1107/S0021889899003532*

[13] Petkov V, Billinge S.J.L, Larsan P, Mahanthi S.D, Vogt T, Rangen K.K, Kanatzidis M.G, Phys. Rev. Vol. B 65 (2002) p. 092105. *http://dx.doi.org/10.1103/PhysRevB.65.092105*

[14] Rehr J.J, Albers R.C, Theoretical approaches to X-ray absorption fine structure. Reviews of Modern Physics vol.72 (2000) p. 621–654. *http://dx.doi.org/10.1103/RevModPhys.72.621*

[15] Stumm von Bordwehr R, A History of the X-ray Absorption Fine Structure, Annales de Physique Vol.14 (1989) pp. 377–465. *http://dx.doi.org/10.1051/anphys:01989001404037700*

[16] Sayers D.E, Stern E.A, Lytle F.W, New Technique for Investigating Noncrystalline Structures: Fourier Analysis of the Extended X-Ray—Absorption Fine Structure. Physical Review Letters Vol.27 (1971) pp. 1204–1207 *http://dx.doi.org/10.1103/PhysRevLett.27.1204*

[17] Thorpe M.F, Levashov V.A, Lei M, Billinge S.J.L, Notes on the analysis of data for pair distribution functions, Ed. By S.J.L. Billinge and M.F.Thorpe, (Kluwer Academic/Plenum Publishers, New York (2002) pp. 105-128. *http://dx.doi.org/10.1007/978-1-4615-0613-3_7*

CHAPTER V

Conclusion

Keywords

Local Structure, Atomic Correlation, Thermal Vibration, Nearest Neighbor Interaction, Figure of Merit

Contents

5.1 INTRODUCTION

The pictorial representation of the internal structure of a solid will be highly useful for the proper selection of technologically important materials such as thermoelectric materials for device applications. Visual understanding and quantitative measurements of the microscopic structure of selected thermoelectric materials is the major objective of this work. Much useful information can be extracted from the X-ray intensity data both in powder form as well as single crystal form using the available advanced computational techniques. The core of the research work completed includes the following.

- Choosing the thermoelectric materials for the study of the electron level properties, (1) magnesium silicide (Mg_2Si) (2) lead telluride (PbTe) (3) bismuth doped with antimony ($Bi_{1-x}Sb_x$ with x=0.2) (4) bismuth telluride (Bi_2Te_3) (5) antimony telluride (Sb_2Te_3) (6) tin telluride doped with germanium (single crystals) ($Sn_{1-x}Ge_xTe$ with x=0.12, 0.25) and (7) indium antimonide single crystals (InSb).
- Recording of high quality X-ray diffraction data both in powder form as well as single crystal form.
- The study of the average structure of the materials using the least square fitting by Rietveld [Rietveld, 1969] technique using the software JANA 2006 [Petříček *et al.*, 2006].
- The charge density distribution analysis using the versatile technique maximum entropy method (MEM) [Collins, 1982].
- The visualization of the electronic charge distribution in three, two and one dimension.
- The extraction of the electron level properties both qualitatively and quantitatively.
- The local structure analysis of the materials using atomic correlation function called pair distribution function (PDF) [Egami and Billinge, 2003] for the powder samples.

5.2 MAGNESIUM SILICIDE (Mg_2Si)

- The cell parameters obtained using Rietveld refinement [Rietveld, 1969] with the help of the software JANA2006 [Petříček *et al.*, 2006] are highly comparable with the reported ones.
- Large thermal vibration parameters are found both for the individual atoms Mg and Si leading to the use of this material for thermoelectric application.
- Thermal vibration parameter of Mg atom is found to be much higher than Si atom indicating the fact that the thermoelectric character is predominant because of Mg atoms than Si atoms.
- The 3D and 2D MEM pictures show an attractive character (metallic) between Mg-Mg and Si-Si atoms and covalent bonding character between Si-Mg atoms. This mixed bonding nature shows this material to be in the slight semi metal-semiconductor region leading to a good thermoelectric character with moderate figure of merit. This prediction is confirmed by earlier reports stating that this is a narrow indirect band gap material with the band gap energy 0.6 eV [Morris *et al.*, 1958; Stella *et al.*, 1964].
- The numerical values of the mid bond electron densities also reveal the above said mixed bonding nature, leading to the conclusion that the material is a better thermoelectric material.
- The nearest neighbor interactions have been analyzed using pair distribution function [Egami and Billinge, 2003]. The mid bond positions measured using the pair distribution function are also highly comparable with the MEM refinement results.

5.3 LEAD TELLURIDE (PbTe)

- The resulting cell parameters match to two decimal places with the reported ones both using unit cell refinement [Holland and Redfern, 1997] and Rietveld [Rietveld, 1969] based JANA 2006 [Petříček *et al.*, 2006] refinements.
- Thermal vibration parameters were reported as large for both lead atom as well as tellurium atom leading to a better thermoelectric behavior of the material.
- Thermal vibration parameter of the lead (Pb) atom is found to be much higher than that of the tellurium (Te) atom concluding the fact that the thermoelectric character is predominant because of the lead atoms than the tellurium atoms.

- The large lattice vibration factor for lead atoms may make the PbTe material a good thermoelectric material.
- The electron density values for both lead (Pb) atoms and tellurium (Te) atoms are reported to be very large leading to the possibility of maximum charge transfer favoring thermoelectric behavior.
- The electron density of the Pb atom is found to be predominant to that of the Te atom, leading to maximum thermoelectric character of the Pb atom. This can be correlated with the lattice thermal vibration parameter also, which is much higher for Pb than the Te atom.
- Both visual and numerical results on the electron density by MEM confirm more ionic nature than covalent nature, though both are coexisting.
- The two dimensional electron density map shows ionic character with more voids all over the plane except at the atomic positions. This void structure may lead to better thermoelectric behavior because it leads to a phonon glass nature.
- The nearest neighbor interactions have been found using the pair distribution function and the mid bond positions determined using the pair distribution function [Egami and Billinge, 2003] are also highly comparable with the MEM refinement results.
- The methodologies like MEM [Collins, 1982] and the PDF analysis are highly useful when systems like PbTe are doped with other metal atoms to increase its figure of merit and efficiency.

5.4 BISMUTH DOPED WITH ANTIMONY (Bi_{80} Sb_{20})

- The resulting cell parameters for both the Bi atoms and the Sb atoms match with the reported ones using Rietveld [Rietveld, 1969] based JANA 2006 [Petříček *et al*, 2006] refinements.
- Iso-electronic doping of the Sb atoms on the atomic site of Bi was confirmed by the Rietveld [Rietveld, 1969] based multiphase analysis.
- The Debye-Waller factor for $Bi_{80}Sb_{20}$ is found to be much greater than that for the individuals, Bi and Sb. This may be the favorable condition for the thermoelectric behavior of Sb doped Bi systems.
- The 3D and 2D visualization gives a clear cut picture for the perfect doping of the Sb atom on the site of the Bi atom and also the enhanced electron density in the

doped system $Bi_{80}Sb_{20}$ compared to the individual atoms Bi and Sb, which leads the system to be a better thermoelectric material.

- The 2D MEM figures give the explicit idea of the reduced open channels (voids) in $Bi_{80}Sb_{20}$ system and not in Bi or Sb systems. Hence, one can suggest the idea to control the thermal lattice vibration by the proper doping of the Sb atoms in Bi lattice. This is the favorable condition to increase the figure of merit by minimizing lattice thermal conductivity.

5.5 BISMUTH TELLURIDE (Bi_2Te_3)

- The cell parameters obtained from the least square refinements based on the Rietveld technique [Rietveld, 1969] match with the ones earlier reported.
- The MEM [Collins, 1982] one dimensional electron density profile confirms two types of Bi atom and Te atom with different maximum electron density at different layers.
- The electron density values for the Bi atom are much higher than that of the Te atom supporting maximum charge transfer by the Bi atom. Hence, the thermoelectric phenomenon in Bi_2Te_3 is predominant due to the presence of Bi atoms.
- The 2D and 3D MEM figures represent the layered and void structure supporting the thermoelectric character of the material.

5.6 ANTIMONY TELLURIDE (Sb_2Te_3)

- The tetragonal structure with accurate cell parameters were obtained with least square refinements method based on the Rietveld technique [Rietveld, 1969].
- The 3D and 2D MEM pictures clearly visualize the valance charge cloud along with the void structure. This may be a favorable condition for this material to become a better thermoelectric material, because it leads to low lattice thermal conductivity.
- Also, the layered structure for this material, which is visualized with 3D and 2D MEM pictures, supports the thermoelectric behavior.
- The MEM [Collins, 1982] electron density value is higher for the Te atom compared to the Sb atom as the atomic number for the Te atom is much higher than of the Sb atom.

- The layered and void structure found in this material also, just like Bi_2Te_3, supports the higher figure of merit for this thermoelectric material.
- The nearest neighbor interactions from the PDF measurements [Egami and Billinge, 2003] are found to be consistent with the reported experimental results.

5.7 TIN TELLURIDE DOPED WITH GERMANIUM ($Sn_{1-x}Ge_xTe$)

- High purity single crystals were grown by a high temperature high vacuum melt growth technique with two germanium doping levels, x=0.12 and 0.25.
- Both Sn and Te atoms gain the property to increase the Debye-Waller factors as the concentration of germanium increases. The increase in lattice distortion may be a favorable condition for this material to become a good thermoelectric material. One can conclude that the optimum doping of germanium will result in a higher Debye–Waller factor and hence favorably increases the figure of merit in germanium doped Sn(Ge)-Te thermoelectric materials.
- Even a small increase in the number of electrons in the system is differentiated in the atomic core as well as in the electron density contour in the 2D and 3D MEM mapping. This concludes the reliability of the MEM [Collins, 1982] technique and mapping.
- The visual and numerical charge density concludes the covalent interaction in both systems.

5.8 INDIUM ANTIMONIDE (InSb)

- Accurate cell parameters were found for the indium antimonide (InSb) single crystals using the least square based Rietveld refinement method [Rietveld, 1969].
- The 2D MEM mapping gives a clear picture of the strong covalent bonding between the indium and the antimony atoms along the bonding direction [111].
- The numerical value of the mid bond electron density, found from the MEM [Collins, 1982] analysis reveals a strong covalent character of this material.
- Heavy charge concentration with localized nature in this material supports this material to be a better thermoelectric material.

The above research work confirms how the fine electronic properties of the thermoelectric materials can be extracted from the X-ray powder diffraction data using the versatile techniques like MEM and PDF. Techniques adopted in this work successfully give the maximum information qualitatively and quantitatively. The other experimental results combined with the above work can give much information for the selection and design of new thermoelectric materials for future applications.

REFERENCES

[1] Collins D, Nature Vol.298 (1982) p. 49-51. *http://dx.doi.org/10.1038/298049a0*

[2] Egami T, Billinge S.J.L, Underneath the Bragg Peaks: Structural Analysis of Complex Material, Oxford University Press, London (2003).

[3] Morris R. G, Redin R. D, Danielson G. C, Phys. Rev. Vol. 109 (1958) p.1909. *http://dx.doi.org/10.1103/PhysRev.109.1909*

[4] Petříček V, Dušek M, Palatinus L, Institute of Physics, Academy of sciences of the Czech republic, Praha (2006).

[5] Rietveld H. M, J. Appl. Cryst. Vol. 2 (1969) pp. 65-71. *http://dx.doi.org/10.1107/S0021889869006558*

[6] Stella A, Lynch D.W, J. Phys. Chem. Solids Vol. 25 (1964) p. 1253. *http://dx.doi.org/10.1016/0022-3697(64)90023-X*

Keyword Index

About the Author

Dr Ramachandran Saravanan, has been associated with the Department of Physics, The Madura College, affiliated with the Madurai Kamaraj University, Madurai, Tamil Nadu, India from the year 2000. He is the head of the Research Centre and PG department of Physics. He worked as a research associate during 1998 at the Institute of Materials Research, Tohoku University, Sendai, Japan and then as a visiting researcher at Centre for Interdisciplinary Research, Tohoku University, Sendai, Japan up to 2000.

Earlier, he was awarded the Senior Research Fellowship by CSIR, New Delhi, India, during Mar. 1991 - Feb.1993; awarded Research Associateship by CSIR, New Delhi, during 1994 – 1997. Then, he was awarded a Research Associateship again by CSIR, New Delhi, during 1997- 1998. Later he was awarded the Matsumae International Foundation Fellowship in1998 (Japan) for doing research at a Japanese Research Institute (not availed by him due to the simultaneous occurrence of other Japanese employment).

He has guided six Ph.D. scholars as of 2016, and about ten researchers are working under his guidance on various research topics in materials science, crystallography and condensed matter physics. He has published around 100 research articles in reputed Journals, mostly International, apart from around 45 presentations in conferences, seminars and symposia. He has also guided around 50 M.Phil. scholars and an equal number of PG students for their projects. He has attracted government funding in India, in the form of Research Projects. He has completed two CSIR (Council of Scientific and Industrial Research, Govt. of India), one UGC (University Grants Commission, India) and one DRDO (Defense Research and Development Organization, India) research projects successfully and is proposing various projects to Government funding agencies like CSIR, UGC and DST.

He has written 3 books in the form of research monographs with details as follows; "Experimental Charge Density - Semiconductors, oxides and fluorides" (ISBN-13: 978-3-8383-8816-8; ISBN-10:3-8383-8816-X), "Experimental Charge Density - Dilute Magnetic Semiconducting (DMS) materials" (ISBN-13: 978-3-8383-9666-8; ISBN-10: 3-8383-9666-9) and "Metal and Alloy Bonding - An Experimental Analysis" (ISBN -13: 978-1-4471-2203-6). He has committed to write several books in the near future.

His expertise includes various experimental activities in crystal growth, materials science, crystallographic, condensed matter physics techniques and tools as in slow evaporation, gel, high temperature melt growth, Bridgman methods, CZ Growth, high vacuum sealing etc. He and his group are familiar with various equipment such as: different types of cameras; Laue, oscillation, powder, precession cameras; Manual 4-

circle X-ray diffractometer, Rigaku 4-circle automatic single crystal diffractometer, AFC-5R and AFC-7R automatic single crystal diffractometers, CAD-4 automatic single crystal diffractometer, crystal pulling instruments, and other crystallographic, material science related instruments. He and his group have sound computational capabilities on different types of computers such as: IBM – PC, Cyber180/830A – Mainframe, SX-4 Supercomputing system – Mainframe. He is familiar with various kind of software related to crystallography and materials science. He has written many computer software programs himself as well. Around twenty of his programs (both DOS and GUI versions) have been included in the SINCRIS software database of the International Union of Crystallography.

www.ingramcontent.com/pod-product-compliance
Lightning Source LLC
Chambersburg PA
CBHW071229210326
41597CB00016B/1997

* 9 7 8 1 9 4 5 2 9 1 0 0 5 *